WORKING WITH

MOLECULAR
CELL
BIOLOGY
Fourth Edition

A STUDY COMPANION

Brian Storrie, Muriel Lederman, Eric A. Wong,
Richard A. Walker, and Glenda Gillaspy
Virginia Polytechnic Institute and State University

W. H. Freeman and Company
New York

To our students, at Virginia Tech
(Virginia Polytechnic Institute and State University)
and elsewhere, and to our teachers

ISBN 0-7167-3604-7

© 2000 by W. H. Freeman and Company

Printed in the United States of America

Second printing, 2001

Contents

Preface

This book and related materials—the end-of-chapter questions ("Testing Yourself on the Concepts" and "MCAT/GRE-Style Questions"), CD-ROM Questions, Instructor's Solutions, and Test Bank—are the result of our cumulative experience in teaching cellular and molecular biology at both the undergraduate and graduate level at Virginia Tech and other Universities. Moreover, it is the outcome of our conviction that the goal of instruction is that students learn. *Working with Molecular Cell Biology: A Study Companion* is an aid to learning. We cannot guarantee to you, the student, that our efforts exactly parallel your individual molecular cell biology course, but we can assure you that an honest effort has been made on our part to reflect the range of knowledge and complexity that is molecular cell biology today. In doing so, we have written a range of short answer and essay questions that cover representative parts of all major sections of *Molecular Cell Biology, Fourth Edition*. We have also written a range of directed analysis questions of specific experimental situations. And, at the end of this volume, you will find the Complete Solutions to the end-of-chapter questions ("Testing Yourself on the Concepts" and "MCAT/GRE-Style Questions"), which appear in *Molecular Cell Biology*.

As aids to you, the student, all questions and experimental situations are clearly grouped by major *Molecular Cell Biology* section heading. Space is provided with the heading to write the lecture date. Step-by-step answers are provided for all questions at the end of each chapter. Answers for short answer and essay questions are cross-referenced to specific sections and pages in *Molecular Cell Biology*. Each Analyzing Experiments question is marked for its level of difficulty: one bullet next to the question denotes appropriateness for sophomore or junior level students, two bullets for junior or senior level, and three for advanced students. In using this system, we freely acknowledge that these markings are intuitive, based on our individual teaching experiences and expectations. We invite your input based on your own reactions. Comments on the appropriateness of the bullet rankings should be directed to our publisher.

This book would not be possible without the input of many. Our thanks go to several people at W. H. Freeman: Executive Editor Sara Tenney for her involvement in the development of the initial framework of the book and in guiding our efforts to an orderly conclusion; Adrie Kornesiewicz, Supplements Editor, for providing us the tools we needed to complete the project; Kate Ahr for pitching in to complete the assembly of this into a volume; Jodi Isman for managing the copy editing; Bill O'Neal for scrutinizing the text; Jessica Olshen as editorial assistant; and Paul Rohloff for overseeing its production. Our thanks go to New City Media, Blacksburg, VA for the skillful preparation of much of the new artwork for *Working with*. Muriel Lederman would like to thank Bill Zabaronick for his computer skills and patience with multiple printouts.

Special thanks go to our families for inspiring us and for their patience in allowing us the time required to complete such a book.

November 1999

Brian Storrie
Muriel Lederman
Eric A. Wong
Richard A. Walker
Glenda Gillaspy

2 Chemical Foundations

PART A: *Chapter Summary*

Molecular cell biology aims at understanding biological organization, that is, the structure and function of organisms and cells, in terms of the properties of individual molecules, such as proteins and nucleic acids. Complex processes such as evolution, development of an organism from a fertilized egg, motion, perception, and thought follow the rules of chemistry and physics. Here we review the important chemical concepts required to comprehend biological organization. The topics discussed are fundamental to the concepts and experiments presented in later chapters.

We review here first the range of topics discussed in the chapter. We begin with covalent bonds, which connect individual atoms in a molecule, and the structures of carbohydrates, which illustrate the importance of small differences in the arrangement of covalent bonds. Next we consider noncovalent bonds, important stabilizing forces between groups of atoms within larger molecules and betweeen different molecules. In particular, emphasis is placed on hydrogen bonds, which determine many of the properties of water. Phospholipids and their association into a bilayer structure, stabilized by multiple noncovalent bonds, are also considered. The next set of questions reviews the concept of chemical equilibrium, focusing on the properties of chemicals in aqueous solution. We then consider various aspects of biochemical energetics, including the central concept of free energy and ATP (adenosine triphosphate) in capturing and transferring energy in cellular metabolism. The final topic centers on the factors that control the rates of chemical reactions and how enzymes accelerate reaction rates. We then progress from review to analysis and consider a range of experimental examples illustrating the principles developed in this chapter.

PART B: *Reviewing Concepts*

2.1 *Covalent Bonds* (lecture date _____)

1. Storage polysaccharides such as glycogen are highly branched and found chiefly in the liver and skeletal muscles. What might be the functional advantages of a branched structure for a storage polysaccharide?

2.2 *Noncovalent Bonds* (lecture date _____)

2. Lipoproteins are solid, spherical complexes of proteins and lipids. They are found in the blood of animals. Their function is to transport nonpolar lipids (e.g., tristearin and other triacylglycerols) and cholesterol esters (cholesterol esterified to a fatty acid) from one place to another in the body. In addition to these nonpolar lipids, some phospholipids are found in lipoproteins. Which of these components, nonpolar lipids or phospholipids, do you think would be on the outside of lipoprotein particles in contact with the (aqueous) blood plasma and which would be on the inside? Explain your answer in terms of the chemical properties of nonpolar lipids and phospholipids.

2.3 *Chemical Equilibrium* (lecture date _____)

3. If 1 ml of a solution of 0.01 M HCl is diluted to 100 ml at 25°C, what is the pH of the resulting solution?

4. If 1 ml of a solution of 1 M NaOH is diluted to 100 L at 25°C, what is the pH of the resulting solution?

5. The pK_a for the dissociation of acetic acid is 4.74. Calculate the molar concentration of acetic acid [HAc] and of sodium acetate [Ac$^-$] you would need to use to make a buffer that is 0.1 M in total acetate concentration and at pH 5.

6. The following reaction is catalyzed by the enzyme phosphoglycerate kinase:

1,3-bisphosphoglycerate + ADP

↔ 3-phosphoglycerate + ATP

$\Delta G°'$ for this reaction is –4.5 kcal/mol. Assuming that the value of [ATP]/[ADP] is 10 and the temperature is 25°C, calculate the concentration ratio of 3-phosphoglycerate to that of 1,3-bisphosphoglycerate at equilibrium.

2.4 Biochemical Energetics
(lecture date _____)

7. A solution of 8 M urea is sometimes used in the isolation of protein molecules. When the solution is prepared by dissolving urea in water at room temperature, it becomes cold. What can you infer about the ΔH, ΔS, and ΔG values for the dissolution of urea in water?

8. Catabolic reactions are involved in the breakdown of cellular fuels, while anabolic reactions are involved in the biosynthesis of cellular materials. Which type of reaction do you think would generally produce ATP? Which type of reactions would consume it?

9. Hydrolysis of ATP to produce ADP under standard conditions has a ΔG value of –7.3 kcal/mol. The hydrolysis of ATP is often used as a source of energy to "pump" substances against a concentration gradient into or out of a cell. Assume that the hydrolysis of one molecule of ATP is coupled to the transport of one molecule of substance A from the inside to the outside of a cell. At 25°C, if the concentration of substance A inside the cell is 100 μM, what is the maximum concentration of substance A outside the cell against which the pump can export it?

10. Ubiquinone, also called coenzyme Q or CoQ, is found in inner mitochondrial membranes, where it serves as an "electron carrier." In this capacity, CoQ undergoes an oxidation-reduction reaction:

$$CoQ + 2e^- + 2H^+ \leftrightarrow CoQH_2 \quad E'_0 = 0.10$$

In another oxidation-reduction reaction, O_2 acts as the ultimate electron acceptor in the mitochondria:

$$1/2 O_2 + 2e^- + 2H^+ \leftrightarrow H_2O \qquad E'_0 = 0.82$$

Calculate the change in electric potential $\Delta E'_0$ and the change in free energy $\Delta G°'$ when $CoQH_2$ is oxidized by O_2 under standard conditions in the following reaction:

$$CoQH_2 + 1/2 O_2 \leftrightarrow CoQ + H_2O$$

2.5 Activation Energy and Reaction Rate
(lecture date _____)

11. Relatively small changes in the free energy of the transition state of a chemical reaction, $\Delta G^\ddagger$, can lead to large changes in the overall rate of the reaction. How can this be?

PART C: Analyzing Experiments

2.1 Covalent Bonds, 2.2 Noncovalent Bonds

12. (•) Lipopolysaccharide (LPS) is a major component of the outer leaflet of the outer membrane of *Escherichia coli* and other gram-negative bacteria. LPS is anchored to the bacterial outer membrane through its lipid A domain. Lipid A contains six fatty acyl chains. As shown in Table 2-1, the composition of the lipid A domain of LPS varies when *E. coli* is grown at 37°C versus 12°C. For reference, the chemical composition of common fatty acids is summarized in *Molecular Cell Biology*, Fourth Edition, p. 27, Table 2-2 and space filling models of example saturated and unsaturated fatty acids are illustrated in MCB, p. 27, Figure 2-18.

a. What is the expected effect of increased palmitoleate levels on the entropic state of the *E. coli* outer membrane?

b. Biological membranes typically have a fluid-like state. What is the expected effect of increased palmitoleate levels on the viscosity of the *E. coli* outer membrane?

c. How does varying the relative levels of laurate and palmitoleate over the temperature range of 12°C to 37°C contribute to maintaining a relatively constant membrane fluidity?

d. Various amino acids, ions, and sugars are transported across lipid bilayers by transport proteins. How may maintaining membrane fluidity be important to the function of transport proteins?

LPS is medically important as endotoxin molecules which produce fever during infections.

	E. coli	
Fatty Acid	Grown at 12°C (μmol/mg LPS)	Grown at 37°C (μmol/mg LPS)
Laurate	0.05	0.16
Palmitoleate	0.10	<0.01

Table 2-1

13. (••) One approach to detecting protein complexes is immunoprecipitation. In this approach, cell extracts are incubated under mild conditions with an antibody directed against one protein. The immunocomplex is then isolated and analyzed for molecular components. Figure 2-1 shows the results of one such experiment using a cell extract containing newly synthesized proteins which had incorporated radioactive methionine. The antibody is directed against p27. The immunoprecipitate is solubilized by detergent treatment and separated into components on the basis of charge in one dimension and molecular weight in the second dimension. The radioactivity is assayed by a phosphoimager and individual proteins appear as dark spots against a white background in the separation media.

a. Why are no dark spots corresponding to the immunoprecipitating antibodies seen in Figure 2-1?

b. You find that one spot, A, is present in the phosphoimager pattern of mock immunoprecipiated cell extracts even though no p27 antibody has been added. What is the likely origin of A?

c. How many protein components appear to be present in the original immunoprecipitation complex?

d. Is the original immunoprecipitation complex held together by covalent or noncovalent bonds?

e. The p24 and p25 spots, respectively, contain 30,000 and 13,000 counts of radioactivity. P24 and p25 contain, respectively, 10 methionines and 4 methionines per protein molecule. What is the relative copy number of p24 and p25 in the complex?

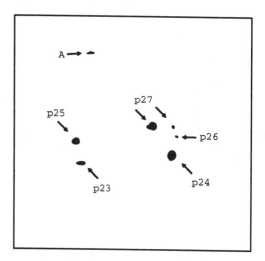

Figure 2-1

2.3 *Chemical Equilibrium,* 2.4 *Biochemical Energetics*

14. (••) Transport proteins are an interesting example of energy utilization in biological systems. Transport proteins are involved in the movement of amino acids, ions, and sugars across biological membranes. One such example is CzcA, a protein which mediates resistance to millimolar concentrations of Co^{2+}, Zn^{2+}, and Cd^{2+}. With varying kinetics, CzcA mediates the efflux of each of these divalent cations from cells and maintains the low intracellular concentration of each. Figure 2-2 shows the dependence of cation transport by CzcA on the concentration of metal ion.

a. What evidence does this provide for cation binding to CzcA?

b. Is the ΔG value for cation transport positive, negative, or zero and why?

c. Cation transport is not directly linked to ATP consumption. Rather CzcA is also a transporter for protons as well as metal cations. The protons are transported in the opposite direction to metal cations. What must be the ΔG value for proton transport for the overall energetics of CzcA to be favorable?

d. Mutation of a negatively charged amino acid to a positively charged amino acid in position 402 of CzcA results in a greatly decreased rate on zinc transport. What is the likely basis of this?

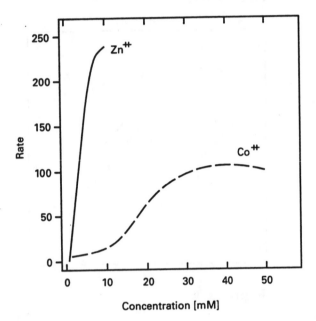

Figure 2-2

Answers

1. Branched polysaccharides have numerous free ends available for formation of glycosidic bonds. Therefore such compounds can rapidly incorporate large amounts of glucose when it is in excess and, conversely, correspondingly rapidly release glucose by hydrolysis when it is in short supply.

For further study see text section "α and β Glycosidic Bonds Link Monosaccharides," p. 21.

2. Phospholipids, which are amphipathic structures, cover the surface of lipoproteins with their polar hydrophilic ends in contact with the blood plasma. Nonpolar lipids, which are hydrophobic, form the cores of lipoprotein particles, where they are separated from the aqueous solution.

For further study, see text section "Phospholipids Are Amphipathic Molecules," p. 26.

3. pH = 4. First calculate the H^+ concentration of the diluted solution and then take the negative log of that value.

$$[H^+] = \frac{(10^{-2}\ M)\ (1\ ml)}{10^2\ ml} = 10^{-4}\ M$$

$$pH = -\log[H^+] = -\log[10^{-4}] = 4$$

For further study, see text section "Biological Fluids Have Characteristic pH Values," p. 31.

4. pH = 9. First calculate the OH^- concentration of the diluted solution; then determine the H^+ concentration and take the negative log of that value.

$$[OH^-] = \frac{(1\ M)\ (1\ ml)}{10^5\ ml} = 10^{-5}\ M$$

$$[H^+][OH^-] = 10^{-14}\ M^2$$

or

$$[H^+] = \frac{10^{-14}\ M^2}{10^{-5}\ M} = 10^{-9}\ M$$

$$pH = -\log[H^+] = -\log[10^{-9}] = 9$$

For further study, see text section "Biological Fluids Have Characteristic pH Values," p. 31.

5. The concentration of sodium acetate $[Ac^-] = 0.0645M$ and of acetic acid $[HAc] = 0.0355M$. These values are calculated using the Henderson-Hasselbach equation (see MCB, p. 32).

$$pH = pK_a + \log \frac{[A^-]}{[HA]} \quad \text{or}$$

$$pH - pK_a = \log \frac{[Ac^-]}{[HAc]}$$

Since $[Ac^-] + [HAc] = 0.1M$, the following substitutions can be made:

$$5 - 4.74 = \log \frac{[Ac^-]}{0.1\,M - [Ac^-]}$$

Taking the antilog of both sides and solving for $[Ac^-]$ gives $[Ac^-] = 0.0645$ M; thus $[HAc] = 0.0355$ M.

For further study see text section "The Henderson-Hasselbach Equation Relates pH and K_{eq} of an Acid-Base System," p. 33.

6. At equilibrium:

[3-phosphoglycerate]/[1,3-bisphosphoglycerate] = 200.

From the definition of the equilibrium constant (see MCB, p. 29),

$$K_{eq} = \frac{[\text{3-phosphoglycerate}][ATP]}{[\text{1,3-bisphosphoglycerate}][ADP]}$$

where brackets indicate equilibrium concentrations. Rearranging the expression $\Delta G^{\circ\prime} = -2.3\,RT \log K_{eq}$ gives

$$\log K_{eq} = \frac{-\Delta G^\circ}{2.3\,RT} \quad \text{(see MCB, p. 38)}$$

Thus,

$$\log K_{eq}$$

$$= \frac{-4500 \text{ cal/mol}}{(2.3)(1.987 \text{ cal/(degree mol)})(298 \text{ K})} = 3.30$$

Taking the antilog of both sides gives $K_{eq} = 2000$. Rearranging from the definition of the equilibrium constant and substituting gives

$$\frac{[\text{3-phosphoglycerate}]}{[\text{1,3-bisphosphoglycerate}]} = K_{eq} \frac{[ADP]}{[ATP]} \quad \text{or}$$

$$(2000)\frac{1}{10} = 200$$

For further study, see text section "Equilibrium Constants Reflect the Extent of a Chemical Reaction," p. 29, text section "The $\Delta G^{\circ\prime}$ of a Reaction Can Be Calculated from Its K_{eq}," p. 38.

7. The fact that the solution becomes cold means that heat is absorbed; that is, the reaction is endothermic, and ΔH for the reaction is positive. Since urea in fact dissolves under these conditions, ΔG must be negative. In order for ΔG to be negative when ΔH is positive, ΔS must be positive. Indeed, the increase in the degree of disorder when urea is dissolved in water is the driving force of the dissolution reaction.

For further study, see text section "The ΔG of a Reaction Depends on Changes in Heat and Entropy," p. 36.

8. Catabolic reactions, such as the breakdown of glucose, produce ATP for the cell. Anabolic reactions, such as the synthesis of proteins, polysaccharides, or nucleic acids, generally consume ATP.

For further study, see text section "ATP Is Used to Fuel Many Cellular Processes," p. 43.

9. The concentration outside could be as high as 22.6 M. The energy available to power the transport of substance A is the 7.3 kcal/mol available from the ATP hydrolysis. The direction of transport described in this question (inside $\rightarrow$ outside of cell) is the reverse of the example given in the text (see MCB, p. 39). To answer the question, you need to calculate the value of C_2 (concentration of A outside the cell) for $\Delta G = +7.3$ kcal/mol and $C_1 = 100$ μM. If the outside concentration were any greater than the value calculated in this manner, the ΔG value for the transport process

would be > +7.3 kcal/mol and, coupled with ATP hydrolysis at −7.3 kcal/mol, ΔG for the overall process would be positive; in this case, the transport of A from inside to outside could not occur.

The ΔG associated with a concentration gradient is given by the expression

$$\Delta G = RT \ln \frac{C_2}{C_1} \qquad \text{(see MCB, p. 37)}$$

where a molecule is being transported from C_1 to C_2; in this case, C_2 = outside concentration and C_1 = inside concentration. Rearranging and substituting into this expression gives

$$\ln \frac{C_2}{C_1} = \frac{\Delta G}{RT}$$

$$= \frac{7300 \text{ cal/mol}}{(1.987 \text{ cal/degree mol})(298 \text{ K})} = 12.3$$

Taking the natural antilog of both sides and solving for C_2 when $C_1 = 100 \, \mu M$,

$$\frac{C_2}{C_1} = 2.26 \times 10^5$$

$$C_2 = (100 \times 10^{-6} \text{ M})(2.26 \times 10^5) = 22.6 \text{ M}$$

Thus under these conditions, the hydrolysis of one mole of ATP would provide energy for one mole of substance A (with an intracellular concentration of 100 μM) to be exported from the cell, as long as the concentration of A outside the cell remained below 22.6 M!

For further study, see text section "Cells Must Expend Energy to Generate Concentration Gradients," p. 39.

10. $\Delta E'_0 = 0.72$ V; $\Delta G^{\circ'} = -33.2$ kcal/mol. The standard electric potential change of an oxidation-reduction reaction is the sum of the E'_0 values of the partial reactions. For the oxidation of $CoQH_2$, the partial reactions are as follows:

$$CoQH_2 \leftrightarrow CoQ + 2e^- + 2H^+ \qquad E'_0 = -0.10 \text{ V}$$

$$\tfrac{1}{2}O_2 + 2\,e^- + 2H^+ \leftrightarrow H_2O \qquad E'_0 = 0.82 \text{ V}$$

Sum: $CoQH_2 + \tfrac{1}{2}O_2 \leftrightarrow CoQ + H_2O \quad \Delta E'_0 = 0.72$ V

The relationship between ΔG and ΔE for an oxidation-reduction reaction is given by the expression (see Table 2-6):

$$\Delta G^{\circ'} \text{ (cal/mol)} = -n\mathscr{F}\Delta E'_0$$

$$= -n \left[\frac{96,500 \text{ joules/(V)(mol)}}{4.18 \text{ joules/cal}} \right] \Delta E'_0 \text{ (volts)}$$

(see MCB, p. 40)

where n is the number of electrons transferred, $\mathscr{F}$ is the Faraday constant, and 4.18 is the factor for converting joules to calories. Substituting $n = 2$ and $\Delta E'_0 = 0.72$ V into this expression and solving, $\Delta G^{\circ'} = -33,200$ cal/mol = −33.2 kcal/mol.

For further study, see text section "The Relationship between Changes in Free Energy and Reduction Potentials," p. 40.

11. The velocity of a reaction may be stated as:

$$V = v \, \frac{[R]}{10^{\Delta G^{\ddagger}/2.3RT}}$$

(rearrangement of last equation MCB, p. 45)

2.3RT = 2.3 (1.987 cal/degree mol)(298 K)

= 1362 cal/mol

Hence every 1362 cal/mol or 1.36 kcal/mol decrease in $\Delta G^{\ddagger}$ results in a full unit decrease in exponent or a ten-fold increase in rate. Small decreases in $\Delta G^{\ddagger}$ do indeed lead to a large increase in rate.

For further study, see text section "Chemical Reactions Proceed through High-Energy Transition States," p. 45.

12a. Because palmitoleate has a cis double double bond which introduces a kink into the structure, it can not pack in a lipid bilayer in as orderly a manner as a saturated fatty acid. Therefore it introduces disorder and increases the entropic state of the *E. coli* outer membrane.

12b. Because of the disorder introduced, increased palmitoleate levels should lead to an increased viscosity at a constant temperature.

12c. As temperature is decreased, increased palmitoleate levels help to maintain normal membrane viscosity.

12d. Transport proteins bind to different amino acids, ions, and sugars. In doing so and effecting transport, they often have transition states that displace portions of the transporter protein molecule with respect to each other. Molecular displacements are facilitated by a viscous membrane.

13a. The phosphoimager assays for radioactivity. The precipitating antibodies are not radioactive.

13b. A is likely to be a pre-existing aggregate. It precipitates without the need for a specific antibody.

13c. In addition to whatever antibody components, there appear to be six additional proteins. Two of these polypeptides have the same molecular weight (27 kDa) as indicated by same Y-axis (molecular weight) position in the two-dimensional gel. They are likely to be different charge forms of the p27 protein with one being a minor form.

13d. As the immunoprecipitate is solubilized by detergent treatment, the precipitate is almost certainly held together by noncovalent bonds.

13e. p24 has approximately 3 times the radioactivity of p25. However, p24 contains 2.5 times as many methionines per protein molecule as p25. Dividing 3 by 2.5 leads to the conclusion that p24 and p25 are approximately present in equimolar amounts in the complex.

14a. The transport velocity for both Zn^{2+} and Co^{2+} saturates with increasing concentration of metal. This strongly indicates that the metal binds to the CzcA protein.

14b. CzcA keeps the intracellular metal concentration low relative to the extracellular concentation. This means moving the cation against an unfavorable concentration gradient. Everything else equal to the ΔG value is likely to be positive.

14c. If the energetics for moving the metal cation are unfavorable, then the ΔG for proton movement must offset this. The negative ΔG for proton movement must have a greater absolute value than the positive ΔG for cation movment.

14d. The divalent cations and protons must be able to bind to CzcA during their transport. The substitution of a positively charged amino acid for a negatively charged one might interfere with binding by charge repulsion.

3

Protein Structure and Function

PART A: *Chapter Summary*

Proteins are the working molecules of a cell. This chapter describes the structure of proteins and how structure gives rise to function. A protein is a linear polymer of amino acids linked together by peptide bonds. The three-dimensional structure of the protein is dictated by its amino acid sequence and is stabilized mainly by noncovalent bonds. Proteins are folded with the assistance of specialized proteins called molecular chaperones and chaperonins. The degradation of many proteins is mediated by the addition of ubiquitin molecules to internal sequences followed by proteolysis in structures called proteosomes.

Enzymes are catalytic proteins that accelerate the rate of cellular reactions by lowering the activation energy. The kinetics of many enzymes are described by the Michaelis-Menten equation. However, some enzymes exhibit allostery, in which the binding of one ligand molecule affects the binding of subsequent ligands. Proteins can be separated by a number of methods based on mass, density, or charge. Some of the common methods include centrifugation, electrophoresis, and liquid chromatography. Centrifugation separates proteins based on their mass and shape. Gel electrophoresis separates proteins based on their rate of movement in an applied electric field. Liquid chromatography separates proteins based on their size, charge, or binding affinity. In addition, protein conformation can be determined using x-ray crystallography, cryoelectron microscopy, and NMR spectroscopy.

Biological membranes contain both integral and peripheral membrane proteins. Integral membrane proteins often span the membrane multiple times. Peripheral membrane proteins are hydrophilic and do not enter the hydrophobic core of the membrane but may be anchored to the cell membrane by covalent linkage to hydrophobic molecules.

PART B: *Reviewing Concepts*

3.1 *Hierarchical Structure of Proteins*
(lecture date _____)

1. What are the four hierarchical levels of protein structure?

2. What is molecular taxonomy?

3.2 *Folding, Modification, and Degradation of Proteins* (lecture date _____)

3. What role do molecular chaperones and chaperonins play in mediating protein folding?

4. What is protein self-splicing?

5. What is the ubiquitin-mediated proteolytic pathway?

3.3 *Functional Design of Proteins*
(lecture date _____)

6. What are the properties of a cellular enzyme?

7. What is feedback inhibition?

8. How can phosphorylation regulate activity of a protein?

3.4 *Membrane Proteins* (lecture date _____)

9. What is the difference between integral and peripheral membrane proteins?

10. How are some cell surface and cytosolic proteins anchored to the membrane?

3.5 *Purifying, Detecting, and Characterizing Proteins* (lecture date _____)

11. What methods can be used to separate proteins based on mass or charge?

12. What is the function of sodium dodecyl sulfate (SDS) during separation of proteins with polyacrylamide gel electrophoresis?

13. What techniques are used to determine the three dimensional structure of proteins?

PART C: *Analyzing Experiments*

3.1 *Hierarchical Structure of Proteins*

14. (••) You have isolated a 31-kDa protein (X) from a bacterium. After producing antibodies to this protein by injecting it into rabbits, you made an affinity column by coupling the anti-X antibodies to a chromatography resin. You applied a homogenate (a mixture obtained by mixing tissue with buffer in a blender) from mouse liver to the column. Upon eluting the bound protein, you performed electrophoresis and identified a 45-kDa protein (Y) as a single band on a SDS-polyacrylamide gel. You performed a similar experiment with a homogenate from bovine (cow) liver and again obtained a 45-kDa protein (Z). Each of the three proteins was treated with a very low concentration of trypsin. No hydrolysis of the bacterial protein occurred, while the mouse and cow proteins each produced fragments of 31 and 14 kDa. Sequence analysis of each of the polypeptides revealed that 48 percent of the amino acids in the bacterial X protein were identical to those in the 31-kDa mouse fragment and

43 percent were identical to those in the 31-kDa cow fragment. Comparison of the mouse and cow tryptic fragments showed that 89 percent of the residues in the 31-kDa fragments were identical, as were 92 percent of the residues in 14-kDa fragments. However, there was no significant homology between the mouse or cow 14-kDa peptides and the bacterial protein.

a. Which two of the three proteins are most closely related?

b. What do the data suggest about the domain structure of the mouse and cow proteins compared with that of the bacterial protein?

c. What do the data suggest about a possible mechanism for the evolution of the mouse and cow proteins?

3.3 *Functional Design of Proteins*

15. (••)Two enzymes, enzyme X and enzyme Z, can both use substrate D to produce product E. The assay results are shown in Table 3-1.

a. Using the data shown in Table 3-1, plot the reaction velocity (μmol E produced/min) versus substrate concentration ([D] mM) for both enzyme X and enzyme Z using the graph shown in Figure 3.1. What type of enzyme kinetics do enzymes X and Z show?

| | Reaction velocity | |
| | μmol E produced/min | |
[D] mM	Enzyme X	Enzyme Z
1.0	1.4	0.2
2.0	2.7	0.4
3.0	3.6	1.3
4.0	4.3	3.7
5.0	4.9	5.5
7.5	5.7	5.9
10.0	6.0	6.0

Table 3-1

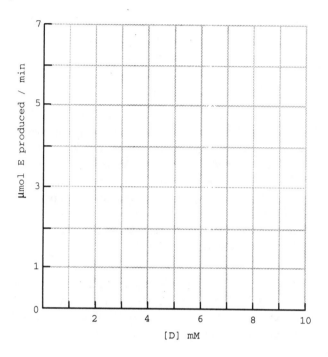

Figure 3-1

b. What is the Km for both enzyme X and enzyme Z for substrate D?

c. How do the activities of enzyme X and enzyme Z differ in the range of 3 to 5 mM D?

d. Based on this data, what can you conclude about the binding of substrate D to enzymes X and Z?

16. (•)Consider the following metabolic pathway:

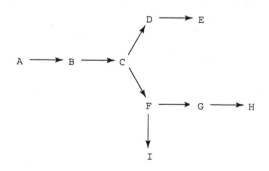

a. Which enzymatic step is likely to be inhibited by the accumulation of product E?

b. Which enzymatic step is likely to be inhibited by the accumulation of product H?

c. What is the likely molecular mechanism by which E and H inhibit the enzymes?

17. (••) Enzyme M acts on substrate A to produce product B. Compound C is an allosteric activator of enzyme M. With genetic techniques to be discussed in later chapters, the valine at position 57 in enzyme M was altered. The altered and unaltered proteins were purified, and their activities were measured in a standard assay system, which included 10 mol A/ml of reaction mixture (a saturating level of A), in the presence and absence of C, the allosteric effector. The assay results are presented in Table 3-2.

	Enzyme activity (nmol B prod./(min)(mg enzyme M))	
Alteration	No C added	1 mM C added
None	10.3	51.4
Val 57 → Ser 57	10.5	30.2
Val 57 → Glu 57	10.2	11.1
Val 57 → Ala 57	10.1	49.5

Table 3-2

a. Is valine-57 more likely to be part of the active site or the allosteric site of enzyme M?

b. In terms of the chemical properties of amino acids, suggest why the substitution of serine, glutamine, and alanine for valine had different effects on the activity of enzyme M.

3.5 Purifying, Detecting, and Characterizing Proteins

18. (••) Protein A and protein B were analyzed using two-dimensional gel electrophoresis as shown in Figure 3-2.

a. What can you conclude about the size and net charge of protein A and protein B at neutral pH?

b. How can you separate protein A from protein B using an ion exchange column that consists of positively charged beads (an anion exchanger)?

c. Using gel filtration chromatography, protein B was found to elute from the column at an apparent molecular weight of 60 kDa. How can you explain the discrepancy in molecular weight of protein B when determined by two-dimensional gel electrophoresis and gel filtration chromatography?

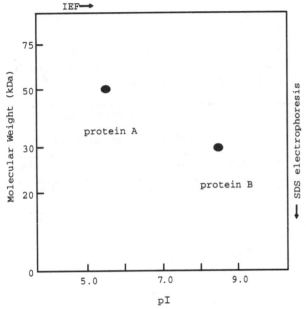

Figure 3-2

similarities in protein structure. In this way evolutionary trees can be constructed based on a comparison of amino acid sequences of selected proteins.

For further study, see text, p. 61.

3. Two general families of chaperones carry out folding of proteins: molecular chaperones and chaperonins. Chaperones bind and stabilize unfolded or partially folded proteins and as a result prevent these proteins from being degraded. Molecular chaperones are thought to bind to all nascent peptides as they are being synthesized on ribosomes. Chaperonins directly facilitate in the folding of proteins.

For further study see Animation 3.2 "Chaperone-mediated Folding" on the MCB 4.0 CD-ROM, as well as text section "Folding of Proteins In Vivo Is Promoted by Chaperones," pp. 63–64 and Figure 3-15.

4. Protein self-splicing is an unusual process found in bacteria and lower eukaryotes in which an internal segment of a polypeptide, an intein, is removed and the ends of the polypeptide are rejoined. This is a process very similar to the removal of introns during the processing of eukaryotic messenger RNAs.

For further study, see text, p. 66 and Figure 3-17.

5. Proteins targeted for degradation by the ubiquitin-mediated pathway are degraded in a two-step process. First, a chain of ubiquitin molecules is added to an internal lysine side chain of the target protein. Additional ubiquitin molecules are added forming a multiubiquitin chain. The ubiquitin-tagged protein is then degraded in a large cylindrical multisubunit complex called a proteosome.

For further study, see Animation 3.1, "Life Cycle of a Protein" on the MCB 4.0 CD-ROM, as well as text section "Cells Degrade Proteins via Several Pathways," p. 67 and Figure 3-18.

6. A cellular enzyme is a protein catalyst that accelerates the rate of a chemical reaction under mild conditions of a cell, such as in an aqueous solvent at 37°C, and pH 6.5–7.5. Enzymes act to increase the rate of a reaction by lowering the activation energy and stabilizing transition state intermediates.

Answers

1. The four hierarchical levels of protein structure are the primary, secondary, tertiary, and quaternary structure. The primary structure is the sequence of amino acid residues. The secondary structure is the localized organization of parts of a polypeptide chain. The tertiary structure is the overall conformation of a polypeptide chain, and the quaternary structure describes the number and relative positions of the subunits of a multimeric protein.

For further study, see text section "Four Levels of Structure Determine the Shape of Protein," p. 54.

2. Molecular taxonomy is the classification of proteins based on similarities and differences in their amino acid sequences. This form of taxonomy provides much information about the function of proteins based on

For further study, see text section "Enzymes Are Highly Efficient and Specific Catalysts," p. 71.

7. Feedback inhibition is a mechanism by which an enzyme that catalyzes an early reaction in a multistep pathway is inhibited by an ultimate product of the pathway. This mechanism is energetically favorable because it prevents both the production of intermediate products and unnecessary metabolic activity.

For further study, see text, p. 84 and Figure 3-30.

8. One of the most common mechanisms for regulating protein activity is the addition and removal of phosphate groups from serine, threonine, or tyrosine residues. Protein kinases catalyze phosphorylation and phosphatases catalyze dephosphorylation. The activity of the target protein cycles between active and inactive depending upon the phosphorylation state of the protein.

For further study, see text section "Cyclic Protein Phosphorylation and Dephosphorylation," p. 77 and Figure 3-30.

9. Integral membrane proteins, also called intrinsic proteins, have one or more segments embedded in the phospholipid bilayer. Most integral membrane proteins contain hydrophobic amino acids which interact with fatty acyl groups of the lipid bilayer. Peripheral membrane proteins, or extrinsic proteins, do not interact with the hydrophobic core of the phospholipid bilayer.

For further study see text section "Proteins Interact with Membranes in Different Ways," p. 78 and Figure 3-32.

10. Some cell surface proteins are anchored to the exoplasmic face of the plasma membrane by a complex glycosylated phospholipid, such as glycosylphosphatidylinositol. Some cytosolic proteins are anchored to the cytosolic face of the membranes by a hydrocarbon moiety, such as prenyl, farnesyl, and geranylgeranyl groups. Other cytosolic proteins are anchored in the membrane by fatty acyl groups such as myristate and palmitate.

For further study, see text section "Covalently Attached Hydrocarbon Chains Anchor Some Proteins to the Membrane," p. 81 and Figure 3-36.

11. Proteins can be separated based on their mass (molecular weight) or charge using a variety of techniques. Centrifugation, gel electrophoresis, gel filtration, and time-of-flight mass spectrometry are four methods used for separating proteins based on their mass. Ion exchange chromatography separates proteins based on their charge.

For further study, see text, pp. 85–90, p. 94, and Figures 3-40, 3-41, 3-42, 3-43, and 3-47.

12. SDS binds to a polypeptide at a ratio of approximately one SDS molecule per amino acid residue. The binding denatures or unfolds a protein so that chain length (i.e., molecular weight) becomes the most important determinant of motion through the pores of the polyacrylamide gel. In other words, the effects of variations in protein shape and charge are minimized.

For further study, see text section "SDS-Polyacrylamide Gel Electrophoresis," p. 87 and Figure 3-41.

13. Three-dimensional structures of proteins are determined by x-ray crystallography, NMR spectroscopy, and cryoelectron microscopy. X-ray crystallography requires protein crystallization. NMR is useful for analyzing small proteins, whereas cryoelectron microscopy is useful for analyzing large protein complexes.

For further study, see text section "Protein Conformation Is Determined by Sophisticated Physical Methods," pp. 95–96.

14a. The mouse and cow proteins are clearly more closely related than either of them is to the bacterial protein. The overall homology between these two proteins is about 90 percent.

14b. The data suggest that the mouse and cow proteins each contain a 14-kDa domain that is not contained in the bacterial protein and that can be hydrolyzed from the rest of the protein by mild trypsin treatment. The larger size of the two mammalian proteins and the lack of homology between their 14-kDa fragment and the bacterial protein support this hypothesis. The existence of an easily accessible trypsin site between the 14-kDa and 31-kDa sections of the mouse and cow proteins also supports the hypothesis that these two regions form structurally distinct domains.

14c. The data suggest that the 31-kDa part of both the mouse protein and cow protein was derived from the same ancestral protein as the bacterial protein, whereas the 14-kDa domain of these mammalian proteins was added at a later date in their evolution. The relatively small differences in the cow and mouse proteins suggest that these proteins shared a more recent common ancestral protein.

15a. A plot of the reaction velocity (μmol E produced/min) versus substrate concentration ([D] mM) for both enzyme X and enzyme Z is shown in Figure 3-3. Enzyme X shows the standard hyperbolic curve characteristic of an enzyme that follows Michaelis-Menten kinetics. In contrast, enzyme Z shows sigmoidal kinetics, indicative of cooperative substrate binding.

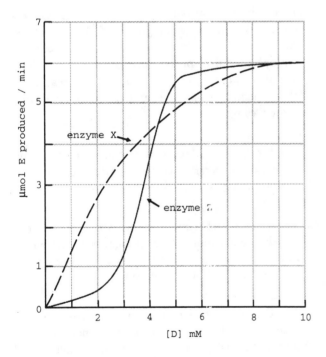

Figure 3-3

15b. Km is a measure of the affinity of the substrate for the enzyme. The Km for an enzyme is defined as the concentration of substrate that gives half maximal velocity (1/2 V_{max}). The V_{max} for both enzymes is 6 μmol E produced/min, however the shapes of the two curves are different. From the curves, it can be estimated that the Km for enzyme X is 2.3 mM D and for enzyme Z is 3.7 mM D.

15c. The shapes of the curves for enzyme X and enzyme Z differ dramatically within the 3–5 mM D range. For enzyme X, the velocity of the reaction increases approximately 36% (from 3.6 to 4.9 μmol E produced/min; see Table 3-1). In contrast, for enzyme Z, the velocity of the reaction increases 4.2-fold or 420% (from 1.3 to 5.5 μmol E produced/min). For the case of enzyme Z, a small change in substrate concentration results in a large change in enzyme activity.

15d. The simple Michaelis-Menten kinetics of enzyme X suggests that substrate D binds to only one site in enzyme X. Whereas, the sigmoidal kinetics of protein Z indicates that there are multiple binding sites for substrate D and binding of substrate D shows cooperative allostery. During cooperative allostery, the binding of one substrate molecule facilitates the binding of subsequent substrate molecules.

16a. Accumulation of product E is likely to inhibit the conversion of C to D, the first step in the pathway that does not lead to the formation of other products.

16b. Accumulation of product H inhibits the conversion of F to G by feedback inhibition.

16c. Products E and H are likely to inhibit the enzymes catalyzing the conversion of C to D and F to G, respectively, by binding to their target enzyme at an allosteric (regulatory) site. Binding at an allosteric site causes the enzyme to assume an inactive conformation, thus inhibiting catalysis at the active site.

17a. The data suggest that valine-57 is part of or related to the allosteric site of enzyme M. Alteration of this residue has little effect on the "basal" activity of enzyme M (the activity in the absence of the activator). The basal activity represents the activity of the active site when the protein's conformation has not been affected by the binding of activator. In contrast, alteration of valine-57 has a large effect on the ability of compound C to act as an activator by binding to the allosteric site. This suggests that replacement of valine-57 either affects binding of the activator or affects the ability of the enzyme to undergo the conformational change that activates catalysis.

17b. Both valine and alanine are nonpolar, uncharged amino acids; thus substitution of alanine for valine is a "conservative" substitution that does not lead to a large functional change. Serine is polar but uncharged, and glutamine is both polar and negatively charged; substitution of each of these amino acids for valine-57 reduces enzyme activity considerably. Thus, in general, the more the chemical properties of an amino acid differ from the residue it replaces, the more drastic its effect on a protein's functional properties.

18a. From the two-dimensional gel, you can determine that the molecular weight of protein A is 50 kilodaltons (kDa) and protein B is 30 kDa. The isoelectric point (pI) is the pH at which the protein has a net charge of zero. The pI for protein A is 5.5.

Therefore at pH 5.5, protein A has a net charge of zero, and at pH 7.0 (neutral pH), protein A would have a net negative charge. Similarly, at pH 7.0, protein B would have a net positive charge.

18b. From 17a, protein A has a net negative charge and protein B has a net positive charge at pH 7.0. Therefore, at neutral pH, protein A will bind to the positively charged beads used in ion exchange chromatography. Protein B will not bind and will flow through the ion exchange column. Protein A can then be eluted from the column using increasing concentrations of salt.

18c. Gel filtration chromatography does not disrupt the subunit structure of a protein, whereas SDS gel electrophoresis separates a protein into its individual subunits. Therefore, we can conclude that protein B normally exists as a dimer of two identical 30 kDa subunits (a homodimer). This would explain why protein B appears to be 60 kDa by gel filtration chromatography but only 30 kDa by SDS gel electrophoresis.

4

Nucleic Acids, the Genetic Code, and Protein Synthesis

PART A: *Chapter Summary*

Geneticists from Gregor Mendel through Thomas Hunt Morgan have studied the hereditary units controlling identifiable traits of an organism. The exact duplication of this information in any species from generation to generation assures the genetic continuity of that species. Chapter 4 focuses on the transmission of this information from the storehouse of deoxyribonucleic acid (DNA) to functioning proteins in the cytosol. Discovery of the structure of DNA in 1953 and the subsequent elucidation of the steps in the synthesis of DNA, RNA, and protein are the monumental achievements of the early days of molecular biology.

To understand how DNA directs synthesis of RNA, which then directs assembly of proteins—the so-called *central dogma* of molecular biology—one needs to be familiar with the structure of nucleic acids. Building on structure, we introduce the general rules of macromolecular carpentry that the cell uses in synthesizing nucleic acids and proteins. Next we outline the roles of mRNA, tRNA, and rRNA in protein synthesis. The chapter closes with a detailed description of the components and biochemical steps in the formation of proteins. Since the events of macromolecular synthesis are so central to all biological functions—growth control, differentiation, and the specialized chemical and physical properties of the cells—they will arise again and again in later chapters. A firm grasp of the fundamentals of DNA, RNA, and protein synthesis is necessary to follow the subsequent discussions without difficulty.

PART B: *Reviewing Concepts*

4.1 *Structure of Nucleic Acids*
(lecture date _____)

1. If the adenine content of DNA from an organism is 36 percent, what is the guanine content?

2. What difference between RNA and DNA helps to explain the greater stability of DNA? What implications does this have for the function of DNA?

3. Which of the common Watson-Crick base pairs in DNA is most stable? Why? How does this property affect the melting temperature of DNA?

4.2 *Synthesis of Biopolymers: Rules of Macromolecular Carpentry*
(lecture date _____)

4. What are four general similarities in the polymeric structure and synthesis of proteins and nucleic acids?

5. How does the enzyme pyrophosphatase participate in DNA replication, in transcription, and in protein synthesis?

4.3 *Nucleic Acid Synthesis*
(lecture date _____)

6. Why is DNA synthesis discontinuous; that is, why is DNA ligase needed to join fragments of one strand of DNA?

7. What are the major differences in the synthesis and structure of prokaryotic and eukaryotic mRNAs?

4.4 *The Three Roles of RNA in Protein Synthesis* (lecture date _____)

8. What is one conclusion that can be drawn from the observation that the genetic code is nearly identical in all cells on earth?

9. Describe two types of experiments used to decipher the genetic code.

10. What is one possible reason why nonstandard base pairing (wobble) is allowed during protein synthesis?

11. What purpose is served by having mRNA, aminoacyl-tRNAs, and various enzymes associated with a large, complicated structure (the ribosome) during protein synthesis?

4.5 *Stepwise Formation of Proteins on Ribosomes*

12. Contrast how selection of the translation start site occurs on prokaryotic, eukaryotic, and viral transcripts.

13. What is the evidence that 23S rRNA in the large ribosomal subunit has a peptidyl transferase activity?

PART C: *Analyzing Experiments*

4.1 *Structure of Nucleic Acids,*
4.3 *Nucleic Acid Synthesis*

14. (•) The compound known as AZT (3-azido-2,3-dideoxythymidine), shown in Figure 4-1, is used to treat patients with acquired immunodeficiency syndrome (AIDS). The effects of AZT treatment vary considerably in different patients, but AZT therapy can result in longer survival times for many AIDS patients. This disease is thought to be caused by the human immunodeficiency virus (HIV), which is a member of the class of viruses

known as retroviruses. Retroviruses contain RNA as their genetic material; a DNA copy of the viral RNA is made during infection by a viral enzyme called *reverse transcriptase.*

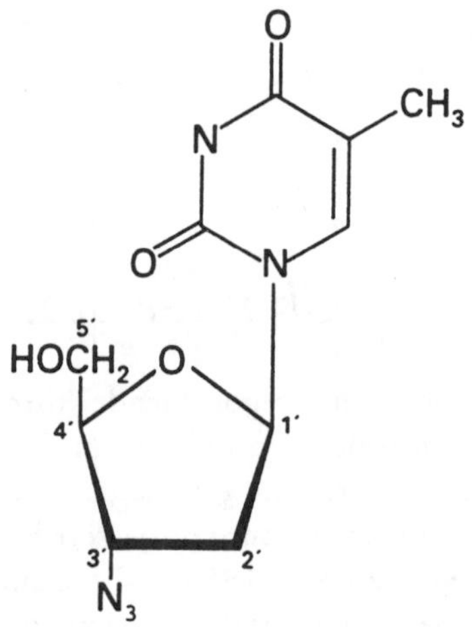

Figure 4-1

a. AZT treatment reduces the amount of HIV present in some patients. What do you think is the mode of action of this drug?

b. Is AZT the active form of the drug; that is, is AZT or some metabolite of AZT responsible for the biological effects in these patients?

c. Long-term treatment with AZT often is associated with the appearance of HIV strains that are resistant to the actions of the drug. What is a likely biochemical or molecular explanation of this observation?

15. (••) High-fidelity replication of DNA is obviously important to all living organisms. Indeed, the error rate for enzymatic synthesis of DNA in eukaryotes is only about one mistake in 10^{10} bases, indicating that cells have developed very accurate synthetic and error-correcting mechanisms. Yet these mechanisms are to no avail if the DNA is damaged after synthesis; such damage can occur if an organism is exposed to radiation, chemicals, or ultraviolet light. Ultraviolet light, which is strongly absorbed by DNA, is in fact lethal to bacteria (Figure 4-2, panel A). However, bacteria can recover from the effects of ultraviolet light in some cases. For example, if UV-irradiated bacteria are exposed to visible light, their percent survival increases compared

with UV-irradiated bacteria kept in the dark (Figure 4-2, panel B). This process is called *photoreactivation;* cells treated in this way are said to be *photoreactivated.* Likewise, if bacteria are irradiated with UV light and then held in the dark in a nonnutritive medium (in which no cell growth occurs) for several hours before the surviving fraction is determined, the percent survival is increased compared with cells assayed immediately after UV irradiation (Figure 4-2, panel C).

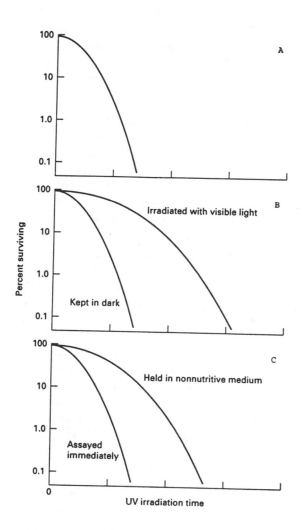

Figure 4-2

a. Why do bacteria with UV-damaged DNA die?

b. What do you think happens to the DNA in photoreactivated cells (panel B)? What is the basis for this opinion?

c. What do you think happens to the DNA in the cells held in the nonnutritive medium (panel C)? Is this the same process that occurs in the photoreactivated cells?

16. The results of irradiation experiments with two bacterial mutants are shown in Figure 4-3. How do these additional data affect your answer to part (c) above?

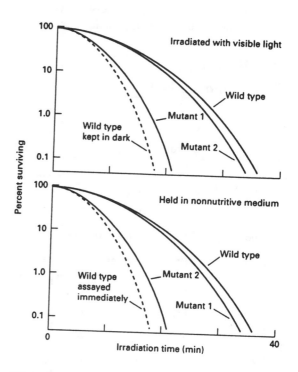

Figure 4-3

Why do you think that bacteria that normally live "where the sun doesn't shine" have evolved these systems for recovering from UV damage?

4.4 The Three Roles of RNA in Protein Synthesis

16. (•)In Table 4-1, representing the genetic code, most of the codons for an individual amino acid are found in the same "box" defined by the 5' nucleotide and in the same column defined by the second nucleotide.

a. What is the explanation for this observation?

b. Methionine and Tryptophan, which are relatively rare amino acids in most proteins, each have only one codon. In addition, the codon for methionine is the start codon, and the codon for tryptophan is in the same box as the stop codons. Both of these codons contain G in the third position and thus are not subject to wobble. In contrast, leucine and serine, which are quite prevalent in many proteins, each have six codons. What do these observations suggest about the evolution of the genetic code?

The genetic code
(RNA to amino acids)*

First position (5' end)	Second position				Third position (3' end)
	U	C	A	G	
U	Phe	Ser	Tyr	Cys	U
	Phe	Ser	Tyr	Cys	C
	Leu	Ser	Stop (och)	Stop	A
	Leu	Ser	Stop (amb)	Trp	G
C	Leu	Pro	His	Arg	U
	Leu	Pro	His	Arg	C
	Leu	Pro	Gln	Arg	A
	Leu	Pro	Gln	Arg	G
A	Ile	Thr	Asn	Ser	U
	Ile	Thr	Asn	Ser	C
	Ile	Thr	Lys	Arg	A
	Met (start)	Thr	Lys	Arg	G
G	Val	Ala	Asp	Gly	U
	Val	Ala	Asp	Gly	C
	Val	Ala	Glu	Gly	A
	Val (Met)	Ala	Glu	Gly	G

• Stop (och) stands for the ochre termination triplet, and Stop (amb) for the amber, named after the bacterial strains in which they were identified. AUG is the most common initiator codon; GUG usually codes for valine, but it can also code for methionine to initiate an mRNA chain.

Table 4-1

4.4 The Three Roles of RNA in Protein Synthesis, 4.5 Stepwise Formation of Proteins on Ribosomes

17. (•••)In the experiments that led to the deciphering of the genetic code, synthetic mRNAs such as polyuridylate were incubated with a cell-free *E. coli* translation system. Although these synthetic polynucleotides were translated slowly (relative to the rates observed for biological mRNAs), the corresponding peptides were produced in sufficient quantity to be analyzed. You are probably wondering why these synthetic mRNAs were ever translated, since they do not contain start codons. The answer lies in the relatively high concentrations of Mg^{2+} (0.02 M) used by Nirenberg and his coworkers in these experiments.

The effects of Mg^{2+} were demonstrated by incubating bacterial ribosomes, a synthetic polyribonucleotide, initiation and elongation factors, tRNAs, and nucleotide triphosphates with 0.005 M Mg^{2+} or 0.02 M Mg^{2+} for 2 min. A portion of each mixture then was centrifuged for 2 h at 100,000g on a 15–40 percent sucrose density gradient. This procedure separates macromolecules on the basis of mass, or S value. After centrifugation, the centrifuge tubes were punctured and the contents allowed to drip slowly into a series of collection tubes. These fractions were assayed for RNA content by measuring the absorbance at 260 nm, a wavelength at which RNA absorbs quite strongly (the bulk of the RNA in these preparations is rRNA). Results of such an analysis, called a *shift assay,* are shown in Figure 4-4. Translation assays also were conducted by adding amino acids to another portion of each incubation mixture and measuring the amount of protein formed. Protein synthesis was observed at the higher magnesium concentration; no protein synthesis could be detected at the lower magnesium concentration.

a. What difference in the interactions among the components in the incubation mixtures with 0.005 M Mg^{2+} and 0.02 M Mg^{2+} is likely to provide the basis for the observed differences in the RNA profiles of the two mixtures?

b. What would be the predicted profile of a similar fractionation performed on a mixture of bacterial ribosomes, initiation and elongation factors, tRNAs, nucleotide triphosphates, and a biologically synthesized mRNA in 0.005 M Mg^{2+}? Why?

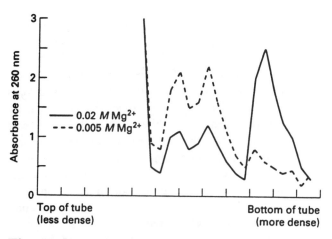

Figure 4-4

4.5 Stepwise Formation of Proteins on Ribosomes

18. (•••) Most eukaryotic mRNAs are translated almost immediately after they are processed and transported to the cytoplasm. A notable exception to this generalization occurs in the eggs of the sea urchin, *Strongylocentrotus purpuratus*. Maternal mRNAs, coding predominantly for histones, tubulin, and actin, are deposited in the cytoplasm of the mature sea urchin egg and serve eventually as the template for almost all protein synthesis for the first 6 h after fertilization. Until fertilization, however, these mRNAs are not translated but remain in the cytoplasm in structures called maternal ribonucleoprotein particles (RNPs). Two general types of hypotheses have been proposed to explain these observations: (1) maternal mRNA in these particles is "masked" and cannot be translated until something is removed, or (2) some component of the translational machinery is missing until after fertilization. Three types of experimental approaches designed to test these hypotheses and determine the parameter(s) that limit protein synthesis in the egg are described below:

(i) Comparison of the protein synthetic capability of cell-free extracts of eggs and of 30-min zygotes (fertilized eggs 30 min after fertilization). Jagus and coworkers have shown that rates of protein synthesis are 10- to 15-fold higher in cell-free extracts of zygotes than in similar extracts of eggs.

(ii) Addition of foreign mRNA to cell-free extracts of sea urchin eggs. Several laboratories have shown that foreign mRNA (e.g., viral mRNA or globin mRNA) is translated in such extracts; however, addition of this mRNA does not cause an increase in total protein synthesis.

(iii) Comparison of the effects of adding sea urchin egg cell-free lysate to an active translational system such as that obtained from the rabbit reticulocyte. (Despite the evolutionary distance between the rabbit and the sea urchin, translation of sea urchin mRNA proceeds normally in the rabbit system.) In Hershey's work, cell-free extracts of unfertilized eggs (A), 30-min zygotes (B), or swimming larvae (pluteus stage, C) from sea urchins were mixed in varying ratios with rabbit reticulocyte lysate; the amount of RNA in each urchin extract was standardized such that equal amounts were added in all the experiments. The rabbit reticulocyte lysate was first treated with a nuclease to destroy any rabbit mRNA, but both lysates should be capable of translating any added mRNA efficiently. The data obtained in these experiments are shown in Figure 4-5. Note that the scale of the ordinate in panel C differs from that in panels A and B.

a. Are data from these three experimental approaches (i–iii) consistent with each other? If so, which hypothesis do these data support? If not, how would you explain the inconsistencies?

b. What additional experiments, using the same materials described in experimental approach (iii), would you design to determine the identity of the components involved in limiting protein synthesis in the sea urchin egg?

c. What result would you expect if you added RNA from eggs or RNA from zygotes to the rabbit reticulocyte lysate and then performed a "shift assay"?

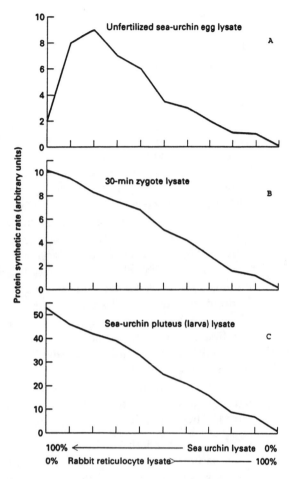

Figure 4-5

4.5 *Stepwise Formation of Proteins on Ribosomes*

19. (•••) Protein synthesis is dependent not only on the various types of RNA but also on a large number of ribosomal proteins. In yeast, for example, there are approximately 73 proteins per ribosome. These proteins are transcribed from widely scattered genes in the yeast genome, and the mRNAs encoding these proteins are translated just like any other mRNA on cytoplasmic ribosomes. The regulatory apparatus controlling ribosomal protein synthesis is complex and incompletely understood at present. It is clear, however, that ribosomal proteins and ribosomal RNAs are synthesized in stoichiometric quantities; that is, cells never accumulate excess ribosomal proteins or excess ribosomal RNA.

Attempts to understand the regulation of this crucial class of proteins have included a series of elegant experiments by Warner's group at Albert Einstein College of Medicine. These workers have taken advantage of the observation that baker's yeast (*Saccharomyces cerevisiae*) can be grown on different media with different growth rates. Yeast grown oxidatively with ethanol as a carbon source have a generation time of approximately 6 h. Yeast grown fermentatively with glucose as a carbon source have a generation time of approximately 2 h and contain more than twice the number of ribosomes per cell than do yeast grown on ethanol. When yeast cultures are shifted from the ethanol medium to the glucose medium ("shift-up"), they exhibit a rapid acceleration of both rRNA and ribosomal protein synthesis. This increased synthesis leads to accumulation of ribosomes until a steady-state level, characteristic of yeast grown on a rich medium, is achieved.

Yeast were grown in synthetic medium containing [^{14}C]leucine for 34 h. Under these conditions all cellular proteins were labeled to steady-state conditions. Before shift-up and at intervals after shift-up to the glucose-containing medium, samples of this culture were incubated for 5 min with [^{3}H]leucine; the radioactive amino acid was then removed. This is called a *pulse-chase* experiment. Cells were then harvested, ribosomal and nonribosomal proteins were isolated, and the ratio of ^{3}H/^{14}C in these proteins was measured. These data were used to calculate the relative rates of synthesis of various proteins (*i*) as follows:

$$A_i = \text{relative rate of synthesis of protein } i$$

$$= \frac{(^{3}\text{H}/^{14}\text{C})_I}{(^{3}\text{H}/^{14}\text{C})\text{total protein}}$$

Results for several proteins are shown in Table 4-2. Data pertaining to the rates of rRNA synthesis in cells at various times after shift-up compared with the rate before shift-up are shown in Figure 4-6.

a. What do these data demonstrate about the co-ordination or lack of coordination of synthesis of different ribosomal proteins?

b. What do these data demonstrate about the co-ordination or lack of coordination of synthesis of ribosomal proteins and rRNA?

c. What changes (if any) would you predict in the relative rates of synthesis of tRNAs before and after a shift-up such as described above?

A^i values at time (min) after shift-up

Protein	0	5	15	30	60	90
Ribosomal proteins:						
2	0.97	1.6	2.03	2.36	2.5	2.65
11	1.09	1.7	2.88	2.62	4.14	3.55
27	0.76	1.41	2.06	2.71	3.54	2.59
61	0.84	1.31	5.53	2.66	3.42	2.68
Average for 13 proteins	.89	1.65	2.29	2.7	3.38	3.01
Nonribosomal proteins:						
A	1.36	1.13	1.33	1.03	0.93	0.88
B	1.16	0.86	0.29	0.59	0.50	0.42
C	1.07	1.09	0.69	1.08	0.94	1.15
D	1.08	0.85	2.05	2.24	2.30	2.25
Average for 4 proteins	1.27	1.11	1.06	1.19	1.32	1.15

Table 4-2

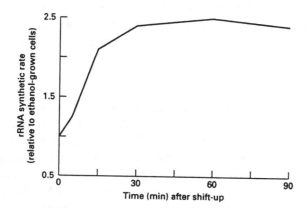

Figure 4-6

Answers

1. Since A = T, the A + T content = 72 percent and G + C content = 28 percent. Since G = C, the content of G is 14 percent.

   For further study, see text section "Native DNA is a Double Helix of Complementary Antiparallel Chains," p. 103.

2. RNA is less stable chemically than DNA because of the presence of a hydroxyl group on C-2 in the ribose moieties in the backbone. Additionally, cytosine (found in both RNA and DNA) may be deaminated to give uracil. If this occurs in DNA, which does not normally contain uracil, the incorrect base is recognized and repaired by cellular enzymes. In contrast, if this deamination occurs in RNA, which normally contains uracil, the base substitution is not corrected. Thus the presence of deoxyribose and thymine make DNA more stable and less subject to spontaneous changes in nucleotide sequence than RNA. These properties might explain the use of DNA as a long-term information-storage molecule.

   For further study, see text section "RNA Molecules Exhibit Varied Conformations and Functions," p. 108.

3. Guanine-cytosine (G-C) base pairs are more stable than adenine-thymine (A-T) pairs, because G and C form three hydrogen bonds, whereas A and T form only two. The greater stability conferred by the additional hydrogen bonds in G-C pairs means that DNA rich in G-C pairs requires more energy for denaturation than does DNA rich in A-T pairs. Thus the melting (denaturation) temperature of G-C-rich DNA is higher than that of A-T-rich DNA.

   For further study, see text section "DNA Can Undergo Reversible Strand Separation," p. 105.

4. (1) Both proteins and nucleic acids are made up of a limited number of subunits (monomers), which are added one at a time, resulting in a linear polymer. (2) The synthesis of both is directed

by a template (mRNA in the case of protein; a complementary strand in the case of nucleic acids). (3) Both are synthesized in one direction only, starting and stopping at specific sites in the template. (4) The primary synthetic product is usually modified; these modifications include cutting, splicing, and addition of chemical groups.

For further study, see text section "Synthesis of Biopolymers," p. 110.

5. Pyrophosphatase catalyzes the breakdown of pyrophosphate (PPi) to two molecules of inorganic phosphate. In DNA replication and in transcription, the α-phosphate of a nucleotide triphosphate is attached to the 3′-hydroxyl of the pentose on the preceding residue, releasing a pyrophosphate. The breakdown of the pyrophosphate by pyrophosphatase is an energetically favorable reaction that helps to drive nucleic acid synthesis. (Synthesis of large polymers from monomers is generally energetically unfavorable due to a large decrease in entropy upon polymer formation.) In protein synthesis, the breakdown of ATP is used to drive the activation of the tRNAs by tRNA-aminoacyl synthetase. ATP is first broken down to AMP as the tRNA is aminoacylated. The reaction is driven further in the direction of tRNA acylation as the released pyrophosphate is broken down by pyrophosphatase. Thus, in all these cases, pyrophosphatase acts to make a synthetic process more energetically favorable by allowing the energy present in both phosphoanhydride bonds of a nucleotide triphosphate to be used to drive an unfavorable reaction.

For further study, see text section "Nucleic Acid Strands Grow in the 5′ to 3′ Direction," p. 112, and "Aminoacyl-tRNA Synthetases Activate Amino Acids by Linking Them to tRNAs," p. 123.

6. DNA synthesis is discontinuous because the double helix consists of two antiparallel strands and DNA polymerase can synthesize DNA only in the 5′to 3′ direction. Thus one strand is synthesized continuously at the growing fork, but the other strand is synthesized in fragments that are joined by DNA ligase.

For further study, see text section "Replication of Duplex DNA Requires Assembly of Many Proteins at a Growing Fork," p. 113.

7. In eukaryotes, the primary transcript resulting from gene transcription must be highly modified to produce a functional mRNA. These modifications, which include cutting, splicing, and modification of the 5′ end (capping) and the 3′ end (polyadenylation), occur in the nucleus before the mRNA is transported to the cytoplasm for translation. In prokaryotes, the primary transcript functions as mRNA without any modification, and translation often begins even before transcription is complete. In addition, many prokaryotic mRNAs are polycistronic (i.e., encode more than one protein), whereas eukaryotic mRNAs are monocistronic (i.e., encode only one protein).

For further study, see text section "Organization of Genes in DNA Differs in Prokaryotes and Eukaryotes," p. 114.

8. A strong conclusion from this observation is that life on earth evolved only once.

For further study, see text section "Messenger RNA Carries Information from DNA in a Three Letter Genetic Code," p. 117.

9. The genetic code was broken by determining which amino acids were present in polypeptides formed by translation of synthetic polynucleotides in bacterial extracts. For example, it was shown that polyuridylate (UUU_n) was translated into polyphenylalanine. Similar analyses of other synthetic nucleotides allowed scientists to discover the triplet codons for all of the standard 20 amino acids.

In a second type of experiment, scientists prepared 20 different bacterial extracts, each containing ribosomes and a mixture of aminoacyl-tRNAs activated with a mixture of 19 unlabeled amino acids and one radioactively labeled amino acid. Each of the 20 extracts contained a different labeled amino acid. Addition of a chemically synthesized trinucleotide that interacted with one of the aminoacyl-tRNAs caused association of that tRNA with ribosomes; the resulting complex was retained on a filter while the free aminoacyl-tRNAs passed through. Detection of radioactivity associated with the filter thus indicated that the trinucleotide present coded for the labeled amino acid in that extract. By mixing each of the 64 possible trinucleotides with a sample of each of the 20 extracts and measuring the radioactivity retained on the filter, scientists determined which amino acid was coded by each trinucleotide. In this way, all 20 amino acids were matched with one or more trinucleotide codons.

For further study, see text section "Experiments with Synthetic mRNAs and Trinucleotides Broke the Genetic Code," p. 119.

10. Wobble may speed up protein synthesis by allowing the use of alternative tRNAs. If only one codon-anticodon pair was permitted for each amino acid insertion, protein synthesis might be temporarily halted until a reasonable level of that particular activated tRNA was regenerated. In fact, as discussed in Chapter 11 of MCB, a slowdown of protein synthesis, due to the lack of a particular aminoacyl tRNA, is used to regulate the levels of enzymes involved in synthesis of some amino acids. This process is called *attenuation*.

For further study, see text section "Nonstandard Base Pairing Often Occurs between Codons and Anticodons," p. 122.

11. The highly specific chemical reactions of translation take place at a much higher rate if the individual components (mRNA, aminoacyl-tRNAs, and the appropriate enzymes) are confined by mutual binding to a common structure, the ribosome. This interaction limits diffusion of one component away from the rest and enables protein synthesis to proceed at the rate of nearly 1 million peptide bonds/s in the average mammalian cell. (Similarly, electron donors and acceptors, in a highly organized array, such as that found in the inner membrane of the mitochondrion or the plasma membrane of a bacterium, can operate much more efficiently than they would if diffusion in three dimensions occurred.) Ribosomes not only provide a site at which the necessary components for protein synthesis are

assembled, but at least one ribosomal component, the (prokaryotic) 23S rRNA, is involved in catalyzing the formation of peptide bonds.

For further study, see text section "Ribosomes Are Protein-Synthesizing Machines," p. 125.

12. In prokaryotes, the eight nucleotide Shine-Dalgarno sequence located near the AUG start codon binds to specific sequences in the 16S rRNA allowing for positioning of the 30S ribosomal subunit near the start site of translation. In eukaryotes, recognition of the start site involves other factors such as *eIF4*, *eIF3* proteins, and Kozak sequences near the start site of the mRNA. *eIF4* recognizes and binds to the 5' cap structure on eukaryotic mRNAs, and *eIF3* proteins are part of the preinitiation complex which is thought to scan along the mRNA, most often stopping at the first AUG. The Kozak sequences facilitate the preinitiation complex in choosing the proper start site.

For further study, see text section "Bacterial Initiation of Protein Synthesis Begins Near a Shine-Dalgarno Sequence in mRNA," p. 129, and "Eukaryotic Initiation of Protein Synthesis Occurs at the 5' End and Internal Sites in mRNA," p. 130.

13. In an assay system 23S rRNA mixed with aminoacylated-tRNAs and peptidyl-tRNA substrates is capable of catalyzing the peptidyl-transferase reaction.

For further study, see text section "During Chain Elongation Each Incoming Aminoacyl-tRNA Moves through Three Ribosomal Sites," p. 131.

14a. AZT is an analog of thymidine, which is a component of DNA. When phosphorylated and converted to AZT triphosphate, it may act as a competitive inhibitor of thymidine triphosphate incorporation during DNA synthesis. An alternative (and more likely) mode of action is as a chain terminator, since AZT does not contain the 3'-OH group needed to form the bond for the addition of the next nucleotide triphosphate. Thus, once AZT is incorporated at the end of a growing DNA strand, it cannot be removed, and DNA synthesis ceases. The unique specificity of

AZT for retroviral infections results from the greater preference of the reverse transcriptase for AZT triphosphate than for thymidine triphosphate; normal human DNA polymerases do not prefer this analog.

14b. As noted in 14a, the active form of the drug is the phosphorylated derivative, not AZT itself.

14c. A likely explanation is that there is a change (mutation) in the reverse transcriptase enzyme lowering its affinity for AZT. If this polymerase is mutated in such a way, then the AZT doses required for viral inhibition will also be inhibitory for the patient's DNA polymerases.

15a. UV irradiation most likely induces changes (mutations) in one or more genes encoding proteins necessary for growth and proliferation. If these changes result in production of nonfunctional proteins (e.g., enzymes), then in time the concentration of functional proteins will decrease due to normal turnover and growth of the bacterial cells will cease. Moreover, if UV-irradiated cells replicate, the daughter cells will carry the mutations in their DNA; thus these cells will produce nonfunctional proteins and will not survive.

15b. The damaged DNA in the photoreactivated cells is repaired before the concentration of functional proteins falls too low for survival. If the DNA were not repaired, the cells would exhibit the same survival curve as the UV-irradiated cells kept in the dark.

15c. The damaged DNA in the cells held in the nonnutritive medium is repaired as well. Since this repair process occurs in the dark, it is probably not exactly the same process that occurs in the photoreactivated cells. The data imply the existence of at least two separate DNA-repair mechanisms.

15d. Mutant 1 is incapable of photoreactivation but can repair damaged DNA if held in a nonnutritive medium for several hours. Mutant 2 is capable of photoreactivation but is incapable of repairing DNA in the dark in a nonnutritive medium. These data further support the hypothesis that at least two DNA-repair systems are present in these bacteria, since components of each can be mutated independently of the other. Enhanced survival of cells maintained in a nonnutritive medium after UV irradiation is called *dark repair* to distinguish it from photoreactivation (light-dependent repair).

15e. The ability to repair UV damage to DNA probably had great survival value in the eons before the earth had an ozone layer (before the appearance of photosynthetic organisms). These repair systems thus reflect the history of the organism. Repair systems, once in place, probably evolved to become capable of repairing other types of damage to DNA. In fact, the enzymes involved in dark repair have been shown to repair DNA damaged by other agents. However, the persistence of photoreactivation in bacteria that live in the dark is currently unexplained, since it seems to repair only thymine dimers, which are characteristic of UV-irradiated DNA.

16a. In most cases, multiple codons for a particular amino acid differ only in the third (3′) base. This occurs because a single tRNA species can bind to multiple codons differing only in the nucleotide found in the third (3′) position. Thus, in Table 4-1 multiple codons for the same amino acid generally appear in the same box and column because their first and second bases are identical.

The formation of "nonstandard," in addition to standard, base pairs between the third base of a codon and the first (5′) base of an anticodon was hypothesized in 1965 by Francis Crick, and is called the wobble hypothesis. Base pairs that can form between codon and anticodon are as follows:

Codon—third (3′) base	Anticodon—first (5′) base
U	A, G, or I
C	G or I
A	U or I
G	C or U

The pairings shown above have several implications. First, codons that differ only by having C or U in the 3′ position are synonymous, since both can base pair with either G or I in the first (5′) position of the anticodon. Second, a codon ending in A-3′ can base pair with anticodons beginning with 5′-I or 5′-U. If the anticodon starts with 5′-I, then codon 5′-(XY)A-3′ will be synonymous with both 5′-(XY)U-3′ and 5′-(XY)C-3′. If the anticodon starts with 5′-U, then 5′-(XY)A-3′ will be synonymous with 5′-(XY)G-3′. Thus, because of synonymous codons, an amino acid with four codons (e.g., Ala), requires only two tRNAs. One will recognize only codons ending in G-3′; the other will recognize 5′-(XY)A-3′, 5′-(XY)C-3′, and 5′-(XY)U-3′. Still, amino acids with only one codon, such as methionine, encoded by (5′)AUG(3′), may be translated unambiguously because anticodons starting with 5′-C will only pair with codons ending in G-3′. These pairing schemes imply that some anticodons are not used; for example, the anticodon (5′)UAU(3′) is not used because it would pair with (5′)AUG(3′), encoding methionine or start, as well as with (5′)AUA(3′), which encodes isoleucine. Likewise, there are no anticodons that pair with the stop codons. In conclusion, those anticodons that are used pair with either one, two, or three codons, depending on the base in the 3′ (wobble) position of the anticodon. If this base is C or A, only one codon is read. If it is G or U, two codons are read. If it is the modified base inosine, then three codons are read.

16b. It is possible that the primitive genetic code, although a triplet code, used only the first two bases of a triplet and that 16 (or fewer) amino acids were actually coded for by primitive replicating entities. The addition of additional amino acids (evolutionary latecomers) would have necessitated the refinement of the code such that the third base of the triplet could be used as part of an unambiguous code. Thus the number of codons for a particular amino acid might be related to the time when that amino acid first was incorporated into the metabolic machinery of living systems; methionine and tryptophan could be evolutionary latecomers.

17a. The analytical technique described separates the 30S (first peak) and 50S (second peak) ribosomal subunits from each other and from fully associated 70S (third peak) prokaryotic ribosomes. At the higher Mg^{2+} concentration, the 30S and 50S subunits form a ternary 70S complex with the synthetic polyribonucleotide, whereas at the lower concentration, they do not (see Figure 4-4). Because this association must occur before protein synthesis is initiated, translation occurs only at the higher Mg^{2+} concentration.

17b. The fractionation profile for a mixture containing a biological mRNA should resemble the profile depicted with the solid line in Figure 4-8; that is, the ribosomal subunits and the mRNA are associated, and protein synthesis can proceed on the 70S ribosomes. This productive association occurs only after the 30S subunit forms a complex with GTP, N-formylmethionyl tRNA, and initiation factors and finds an AUG codon on the biological mRNA. The complex of the 30S subunit, GTP, initiation factors, N-formylmethionyl tRNA, and mRNA can form at low Mg^{2+} concentrations if an AUG triplet is present on the ribonucleotide to be translated.

18a. These data are consistent with each other and do not rule out either hypothesis. Two pieces of evidence are consistent with hypothesis 1 ("masked" RNA): the observation that foreign mRNA can be translated by egg lysates, and the minimal stimulation of the reticulocyte lysate by 10 percent urchin egg lysate (panel A), even though the amount of egg mRNA added would be expected to produce a large stimulation of protein synthesis in the reticulocyte lysate assay system. In contrast, the rate of protein synthesis was much higher when reticulocyte lysate and a 10 percent sea urchin larval lysate were mixed (panel C). Evidence consistent with hypothesis 2 (missing translational component) includes stimulation of translation by addition of reticulocyte lysate (panel A) to sea urchin egg mRNA; clearly this mRNA is not completely "masked," since it can be translated under these conditions. The decline in synthetic rate (left side of panel A) at very low reticulocyte to egg lysate ratios indicates that translation is limited by some component of the reticulocyte lysate. This decline is not seen with the zygote lysates (panel B). The missing component cannot be mRNA, as the reticulocyte lysate had been treated to remove mRNA.

The data are consistent with the hypothesis that some initiation or elongation factor supplied by the reticulocyte lysate becomes rate-limiting in mixtures containing high levels of egg lysate. It seems likely that both mRNA availability and translational factors are limited in the sea urchin egg; that is, both hypotheses are probably true.

18b. You could add purified initiation factors, elongation factors, tRNAs, or ribosomal subunits to the egg lysate and determine the effect of these components on translation of the egg mRNA species. Alternatively, you could fractionate the rabbit reticulocyte lysate using various chromatographic techniques, determine the effects of various fractions on the protein synthetic rate of the egg lysate, and attempt to identify the components in the most active fraction(s). Application of the former approach by Hershey and coworkers implicated the elongation factor eIF_{4F} as the component that limits translation in the egg lysate but not in the zygote lysate.

18c. RNA from eggs should not shift the 40S preinitiation complex into the 80S initiation complex; RNA from zygotes should shift the 40S complex to the 80S complex in this assay.

19a. Synthesis of the 13 ribosomal proteins examined seems to be coordinated, since all of these proteins show an increase in the relative rate of synthesis after shift-up. Furthermore, the time course of this increased synthesis is similar for all the proteins examined.

19b. Synthesis of ribosomal proteins and rRNA seems to be coordinated, since both show an increased rate of synthesis after shift-up. Furthermore, the time courses for the increased synthetic activities are similar for rRNA and ribosomal proteins.

19c. These data imply that at least two of the components of the translation machinery, ribosomal proteins and rRNA, are coordinately regulated. It seems likely that tRNA (and aminoacyl-tRNA synthetases), the other major component of the protein synthetic machinery, also would show a concomitant increase in synthetic rate after shift-up.

5

Biomembranes and the Subcellular Organization of Eukaryotic Cells

PART A: *Chapter Summary*

At both the molecular and cellular level, structure and function are intimately related. This chapter focuses on the basic structures of eukaryotic cells and the methods used to identify and characterize cellular structures. Much of our knowledge of cell architecture is derived from direct observations of cells and subcellular structures by light and electron microscopy. Another basic approach to the study of cell structure and function is to break open cells and separate them into subcellular components for further characterization. Together, microscopy and cell fractionation techniques have provided clear evidence for the universality of biomembrane structure in all cells, and have helped define the specific structure and function of the plasma membrane and each of the organelles present in eukaryotic cells.

PART B: *Reviewing Concepts*

5.1 *Microscopy and Cell Architecture*
(lecture date _____)

1. Based on what you know about the differences between light microscopy and transmission electron microscopy, indicate which technique would be best for visualizing each structure or phenomenon listed below and why. If light microscopy is best, which particular technique (brightfield, phase-contrast, fluorescence, etc.) would be most useful?

   a. Motion of chloroplasts in plant cells
   b. Viral particles
   c. Motion of bacterial cells
   d. Cells containing a specific protein in a tissue such as brain

2. The ability of a microscope to discriminate between two objects separated by a distance D is dependent upon the wavelength of the radiation (λ), the numerical aperture of the optical apparatus (a), and the refractive index of the medium between the specimen and the objective lens (N), according to the following equation:

$$D = (0.61\,\lambda) \div (N \sin a) \quad \text{(See MCB, p. 141)}$$

   a. If you are viewing an object with visible light at a wavelength of 600 nm, in an instrument with a numerical aperture of 70°, what would be the resolution if air (refractive index = 1) was the medium between the specimen and the objective?

   b. What would be the resolution if immersion oil with a refractive index of 1.5 were placed between the specimen and the objective?

   c. What would be the resolution if you used the same immersion oil and viewed the specimen in blue light (wavelength of 450 nm)?

   d. Under any of the conditions specified above, could you see a mitochondrion (average size $1 \times 2\ \mu m$)?

3. In some electron microscopy methods, the specimen itself is not directly imaged. How do these methods provide structural information about cellular structure, and what types of structures may be visualized by these methods?

4. Why must special techniques be used to image living cells? What techniques are used to image structures in thick cells and tissues?

5.2 *Purification of Cells and Their Parts*
(lecture date _____)

5. In terms of the basis for separation of particles, what is the major difference between differential-velocity centrifugation and equilibrium density-gradient centrifugation?

6. Techniques used in the fractionation of cells to obtain preparations of subcellular structures (organelles and plasma membranes) include differential-velocity centrifugation, equilibrium density-gradient centrifugation, binding to antibody-coated beads, and sonication. In what order would these techniques normally be used during cell fractionation and why?

5.3 *Biomembranes: Structural Organization and Basic Functions* (lecture date _____)

7. Although both faces of a biomembrane are composed of the same general types of macromolecules (principally phospholipids and proteins), the two faces are not identical. What accounts for the differences or asymmetry between the faces?

8. What types of experimental evidence support the fluid mosaic model of biomembranes?

9. What factors determine the fluidity of a biomembrane?

5.4 *Organelles of the Eukaryotic Cell*
(lecture date _____)

10. Give two reasons why lysosomal enzymes do not degrade macromolecules located in the cytosol or nucleus of intact cells.

11. What is the function of the vacuole present in plant cells?

12. A major difference between prokaryotes and eukaryotes is the presence of organelles in the latter. Since we think that eukaryotes are more advanced than prokaryotes, a logical conclusion is that the evolution of organelles must confer some advantage(s) to eukaryotes. Describe at least two advantages of organelle structures.

13. What are the functions of the cytoskeleton?

14. Just as the evolution of organelles allows separation of diverse biochemical processes into specific compartments, the evolution of multicellular organisms allows particular cell types to specialize in different activities. This cellular specialization is often accompanied by organelle-specific specializa-tion; that is, cells optimized for a particular role often have an abundance of the particular organelle or organelles involved in that role. Based on what you know about organelle functions, which organelle(s) or membrane(s) would you predict would be over-represented in each of the following cell types?

a. Osteoclast (involved in degradation of bone tissue)

b. Anterior pituitary cell (involved in secretion of peptide hormones)

c. Palisade cell of leaf (involved in photosynthesis)

d. Brown adipocyte (involved in lipid storage and metabolism, as well as thermogenesis)

e. Ceruminous gland cell (involved in secreting earwax, which is mostly lipid)

f. Schwann cell (involved in making myelin, a membranous structure that envelops nerve axons)

g. Intestinal brush border cell (involved in absorption of food materials from gut)

h. Leydig cell of testis (involved in production of male sex steroids, which are oxygenated derivatives of cholesterol)

PART C: *Analyzing Experiments*

5.2 *Purification of Cells and Their Parts*

15. (••) A fluorescence-activated cell sorter (FACS) can be used to identify and purify cells that have a specific fluorescent antibody bound to them. In addition, this instrument can also identify and purify cells with varying amounts of DNA. For example, cells that have just divided and have X amount of DNA can be separated from cells that have duplicated their DNA (2X) and are preparing to divide again. This is done by incubating the cells in a fluorescent dye that binds strongly to DNA.

Because the fluorescence of this dye is directly proportional to the DNA content, cells containing 2X DNA are twice as fluorescent as those containing only 1X DNA. Data from two such analyses are shown in Figure 5-1. The x-axis (channel number) indicates the level of fluorescence of a given cell; a higher channel number means that more fluorescence was measured. The y-axis indicates the number of cells with a given fluorescence level.

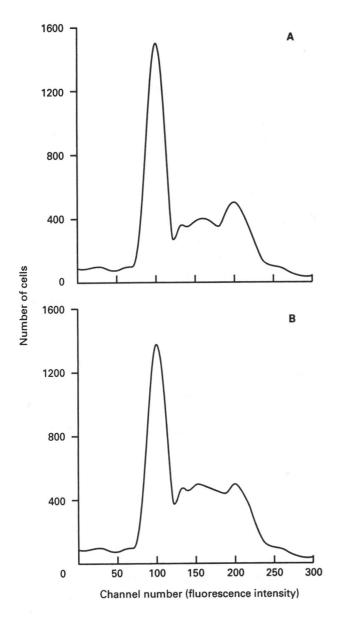

Figure 5-1

a. In panels A and B of Figure 5-1, which portions of the graph correspond to the following cells: G_1 cells, which have not replicated their DNA (1X DNA); G_2 and M phase cells, which have replicated but not segregated their DNA (2X DNA); and S phase cells that

are in the process of replicating? See MCB Figure 13-1, p. 496 for reference.

b. Were the cells used in analysis A (panel A) dividing more or less rapidly than those used in analysis B (panel B)? Explain your answer.

c. Certain drugs can stop or arrest cells in specific phases of the cell cycle. How would the pattern shown in panel A change if the cells had been treated with a drug that arrested cells prior to DNA replication. How would the pattern change if the cells had been treated with a drug that arrested cells in the middle of mitosis. Explain your answers.

5.2 *Purification of Cells and Their Parts,*
5.4 *Organelles of the Eukaryotic Cell*

16. (●●) Mouse liver cells were homogenized and the homogenate was subjected to equilibrium density-gradient centrifugation, using sucrose gradients. Fractions obtained from these gradients were assayed for *marker molecules* (i.e., molecules that are limited to specific organelles). Results of these assays are shown in Figure 5-2. The marker molecules have the following functions: cytochrome oxidase is an enzyme involved in the process by which ATP is formed in the complete aerobic degradation of glucose or fatty acids; ribosomal RNA forms part of the protein-synthesizing ribosomes; catalase catalyzes decomposition of hydrogen peroxide; acid phosphatase hydrolyzes monophosphoric esters at acid pH; cytidylyl transferase is involved in phospholipid biosynthesis; and amino acid permease aids in transport of amino acids across membranes.

a. Indicate the marker molecule for each organelle listed in Table 5-1 and the number of the fraction that is *most* enriched for each organelle.

b. Is the rough endoplasmic reticulum more or less dense than the smooth endoplasmic reticulum?

c. Is the plasma membrane fraction the most dense or the least dense? Why is this the case?

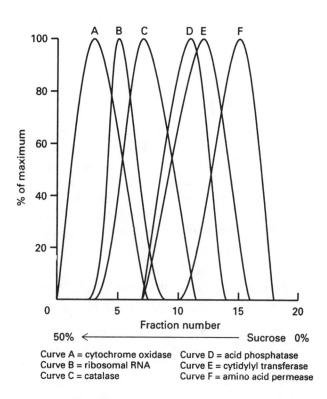

Curve A = cytochrome oxidase   Curve D = acid phosphatase
Curve B = ribosomal RNA   Curve E = cytidylyl transferase
Curve C = catalase   Curve F = amino acid permease

Figure 5-2

Organelle	Marker molecule	Enriched fraction (no.)
Lysosomes	_____	_____
Peroxisomes	_____	_____
Mitochondria	_____	_____
Plasma membrane	_____	_____
Rough endoplasmic reticulum	_____	_____
Smooth endoplasmic reticulum	_____	_____

Table 5-1

5.4 *Organelles of the Eukaryotic Cell*

17. (••) Mammalian cells grown in culture usually contain a representative population of organelles such as mitochondria, Golgi vesicles, lysosomes, etc. These cells can be used as a model system in which to study the formation and dynamics of organelles. In one such study, it was found that cultured hamster cells grown in the presence of 0.03 M sucrose accumulated numerous refractile (very bright in the phase-contrast microscope), sucrose-containing vacuolar structures called *sucrosomes*. Cytochemical staining techniques demonstrated

the presence of the enzyme acid phosphatase in these structures, indicating that they were derived from lysosomes. The structures persisted for many days in the cells but apparently had no ill effects on cellular metabolism.

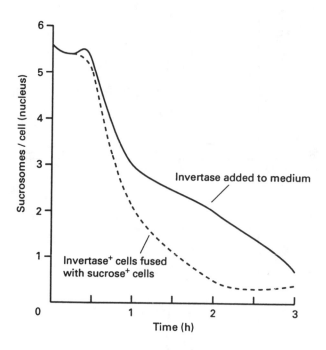

Figure 5-3

a. In one experiment, hamster cells first were grown in the presence of sucrose, and then the enzyme yeast invertase, which catalyzes the cleavage of sucrose into its monosaccharide components, was added to the medium. The number of sucrosomes per cell was monitored following addition of invertase by phase-contrast microscopy, as shown by the solid line in Figure 5-3. Explain why these data indicate that invertase is internalized by these cells. What can you conclude about its subcellular localization? What additional experiments could you perform in order to test this hypothesis?

b. In a second type of experiment, hamster cells grown in the presence of sucrose (sucrose+ cells) were mixed with hamster cells grown in the presence of invertase (invertase+). The two cell types were fused together by the addition of a fusogenic agent, which produces hybrid cells containing two or more nuclei [see MCB, Figure 6-8, p. 189]. During the course of cell fusion, the number of sucrosomes per cell (actually sucrosomes per nucleus) again was monitored by phase-contrast microscopy, as shown by the dashed line in Figure 5-3. From these data, what can you conclude about the subcellular localization of invertase in the fused cells? Are these data consistent with your hypothesis regarding the fate of invertase in unfused cells? Do the results of this experiment suggest that there is a dynamic exchange of molecules between subcellular organelles?

Answers

1a. Motion of chloroplasts in plant cells is best visualized by light microscopy; electron microscopy (EM) is not suitable because EM specimens need to be killed and fixed before they can be viewed. Nomarski interference microscopy and perhaps standard transmission light microscopy would be suitable. However, phase-contrast microscopy is the preferred technique because objects the size of chloroplasts can be easily distinguished using this technique and because the difference in refractive index of the cytosol and the organelles is sufficient to give good contrast. Another suitable technique utilizes the natural fluorescence of the chlorophyll in chloroplasts as a visual marker; in a fluorescence microscope under the proper conditions, the chloroplasts would have a red fluorescence and could be easily visualized.

1b. Viral particles are best visualized by transmission electron microscopy; they are usually too small to be seen with the light microscope.

1c. Motion of bacterial cells could be observed with a light microscope using phase-contrast microscopy. Another light microscopic technique, called dark-field microscopy, would be particularly suitable because this technique allows the use of live, unstained cells and is sufficiently sensitive so that objects the size of bacterial cells are easily resolved. Electron microscopy could not be used because it requires fixed dead cells.

1d. The localization of a specific protein in a tissue is best accomplished with the light microscopic technique known as immunofluorescence microscopy, although the use of a specific antibody attached to an enzyme that generates a colored product also would allow the use of standard bright-field micro-

scopic techniques. Electron microscopy could not be used because the size of the subject (a whole tissue) is prohibitively large. Immunofluorescence and immunocytochemistry are the only techniques that allow specific macromolecules to be localized in tissue sections.

For further study, see text section "Microscopy and Cell Architecture," p. 140.

2a.
$$D = \frac{0.61 \times 600 \text{ nm}}{1 \times \sin 70°}$$
$$= \frac{366 \text{ nm}}{0.94}$$
$$= 390 \text{ nm or } 0.39 \text{ } \mu\text{m}$$

2b. 260 nm or 0.26 μm

2c. 190 nm or 0.19 μm

2d. A mitochondrion should be visible under all of the conditions in 2a–c, since its dimensions are considerably larger than the value of D.

For further study, see text section "Light Microscopy Can Distinguish Objects Separated by 0.2 μm or More," p. 140.

3. Certain electron microscopy methods rely on the use of metal to coat the specimen. The metal coating acts as a replica of the specimen, and the replica rather than the specimen itself is viewed in the electron microscope. Methods that use this approach include metal shadowing, freeze fracturing, and freeze etching. Metal shadowing allows visualization of viruses, cytoskeletal fibers, and even individual proteins, while freeze fracturing and freeze etching allow visualization of membrane leaflets and internal cellular structures.

For further study, see text sections "Preparation of Fixed, Stained Samples for TEM," p. 149, and "Membrane Leaflets Can Be Separated and Each Face Viewed Individually," p. 165.

4. Cellular structures do not typically absorb visible light and therefore provide little contrast when viewed with a basic light microscope.

However, cellular structures do affect the wave properties of light through the phenomena of refraction and diffraction, and techniques such as phase contrast and differential interference contrast light microscopy use these effects to produce contrast from otherwise transparent cells. Thick cells and tissues are difficult to visualize due to image distortion from out-of-focus structures and because it is difficult to get penetration of stains into thick samples. Typically, this meant that samples had to be fixed and sectioned prior to observation. Differential interference contrast light microscopy limits out-of-focus distortion to some extent in unstained specimens, but confocal scanning and deconvolution techniques allow visualization of fluorescent structures in living or fixed thick cells and tissues with little out-of-focus contributions. However, visualization by electron microscopy still requires specimens to be sectioned and stained.

For further study, see text sections "Samples for Light Microscopy Usually Are Fixed, Sectioned, and Stained," p. 141, and "Confocal Scanning and Deconvolution Microscopy Provide Sharper Images of Three-Dimensional Objects," p. 144.

5. Differential-velocity centrifugation separates particles on the basis of their mass (size), while equilibrium density-gradient centrifugation separates particles on the basis of their density.

For further study, see Figures 5-23 and 5-24 in text, p. 155.

6. The order of use for these techniques would be: sonication, differential-velocity centrifugation equilibrium density-gradient centrifugation, and binding of specific structures or molecules to antibody-coated beads. Sonication is used to break open the plasma membrane to allow access to organelles and cytosolic components. Differential-velocity centrifugation of the disrupted cells provides partial fractionation of cell components and subsequent equilibrium density-gradient centrifugation of selected fractions (enriched for the component of interest) would further improve the purity of the component. Finally, use of antibody-coated beads where the antibody is specific to a certain protein (e.g., on the surface of a specific organelle) would produce a highly-purified component.

For further study see text section "Purification of Cells and their Parts," p. 152.

7. Since biomembranes form closed compartments, one face of the bilayer is automatically exposed to the interior of the compartment while the other is exposed to the exterior of the compartment. The two faces therefore face different environments and perform different functions. The different functions are in turn directly dependent on specific molecular composition of each face. For example, different types of phospholipids and integral proteins are typically present on the two faces. In addition, different domains of transmembrane proteins are exposed on each face of the bilayer. Finally, in the case of the plasma membrane, the lipids and proteins of the exoplasmic face are often modified with carbohydrates.

For further study see text section "All Integral Proteins and Glycolipids Bind Asymmetrically to the Lipid Bilayer," p. 162, and "The Phospholipid Composition Differs in Two Membrane Leaflets," p. 162.

8. Two major types of experimental evidence support the fluid mosaic model. The first is based on cell fusion experiments and is a qualitative approach that demonstrates the mixing of plasma membrane components. The second approach, fluorescent recovery after photobleaching (FRAP) allows quantitative measurement of the extent of membrane fluidity and even reveals the movements of specific molecules.

For further study, see text section "Most Lipids and Integral Proteins Are Laterally Mobile in Biomembranes," p. 162.

9. Since the interactions that form the bilayer are noncovalent, temperature is an important influence on the fluidity of biomembranes. Biomembranes have a melting temperature at which a phase transition between a gel-like state and a fluid-like state occurs. In addition to tempera-

ture, the components that comprise the biomembrane may influence fluidity. Phospholipids with shorter fatty acyl chains or that have more unsaturated bonds allow fluidity to be maintained at lower temperatures. Membrane cholesterol also may aid in maintaining fluidity at lower temperatures.

For further study, see text section "Fluidity of Membranes depends on Temperature and Composition," p. 164.

10. The lysosomal membrane keeps the lysosomal enzymes separated from the cytosol and nucleus. In addition, the pH of the cytosol (approx. 7.0) is inhibitory for most lysosomal enzymes, which function best at pH 4–5.

For further study, see text section "Lysosomes Are Acidic Organelles That Contain a Battery of Degradative Enzymes," p. 169.

11. The vacuole in plant cells is a storage site for many small molecules and ions, which are present in sufficient concentrations to result in the movement of water from the cytoplasm into the vacuole. This movement expands the vacuole, creating hydrostatic pressure (turgor) inside the cell. The cell wall acts to counter this pressure. Turgor causes the internal volume of the cell to expand rapidly as cells elongate during plant growth.

For further study, see text section "Plant Molecules Store Small Molecules and Enable the Cell to Elongate Rapidly," p. 170.

12. The presence of organelle structures in cells permits the following advantageous phenomena to occur: (1) compartmentalization of antagonistic processes (e.g., protein synthesis in the endoplasmic reticulum and protein degradation in lysosomes); (2) allocation of membrane-dependent processes to increased intracellular membrane surfaces; and (3) confinement of diffusion-limited processes to a small area, thus increasing the rate of these processes.

For further study see text section "Organelles of the Eukaryotic Cell," p. 168.

13. The cytoskeleton functions to provide structural support and to aid in cell motility.

For further study see text section "The Cytosol Contains Many Particles and Cytoskeletal Fibers," p. 175.

14a. Osteoclasts contain an above-average amount of lysosomal enzymes. In a strict sense, these cells do not have more lysosomes, as the degradative activity occurs outside the cell; osteoclasts could be considered to have a large external lysosome.

14b. Anterior pituitary cells contain an above-average amount of rough endoplasmic reticulum (ER) and Golgi complex; these organelles are involved in synthesis and processing of secretory proteins.

14c. Palisade cells of the leaf contain an above-average amount of chloroplasts, which are solely responsible for photosynthesis in eukaryotes.

14d. Brown adipocytes contain an above-average amount of peroxisomes, which are involved in fatty acid degradation and heat production. In addition, these cells contain above-average amounts of the other oxygen-utilizing, fatty acid–catabolizing organelles, the mitochondria.

14e. Ceruminous gland cells contain above-average amounts of smooth ER, which is the site of lipid biosynthesis.

14f. Schwann cells contain above-average amounts of both smooth and rough ER; these organelles are the site of biosynthesis of the proteins and lipids that compose the myelin sheath.

14g. Intestinal brush border cells contain an above-average amount of plasma membrane (the microvilli), which is the site of nutrient uptake.

14h. Leydig cells of the testis contain above-average amounts of smooth ER, which is the site of cholesterol biosynthesis.

For further study, see text section "Organelles of the Eukaryotic Cell," p. 168.

15a. G₁ cells with 1X DNA are represented by the large peak centered on channel number 100, while G_2 and M phase cells with 2X DNA are represented by the smaller peak centered on channel number 200. S phase cells in the process of replicating are represented in the intermediate channel numbers (120–180).

15b. Comparison of A and B indicates that the cells used in A were dividing more slowly; that is, a smaller proportion were synthesizing DNA (channel numbers 120–180) and a larger number were in the G_1 and G_2/M phases of the cell cycle.

15c. Cells arrested prior to DNA synthesis would be in the G_1 phase of the cell cycle and would have 1X DNA. Analysis with a FACS machine would reveal essentially all cells in a single peak centered around channel number 100. Cells arrested in M phase, which follows DNA replication, would have 2X DNA, and would produce a single peak centered around channel number 200.

For further study, see text section "Flow Cytometry Separates Different Cell Types," p. 153.

16a.

Organelle	Marker molecule	Enriched fraction (no.)
Lysosomes	Acid phosphatase	11
Peroxisomes	Catalase	7
Mitochondria	Cytochrome oxidase	3
Plasma membrane	Amino acid permease	15
Rough endoplasmic reticulum	Ribosomal RNA	5
Smooth endoplasmic reticulum	Cytidylyl transferase	12

16b. The rough endoplasmic reticulum is more dense than the smooth endoplasmic reticulum, since it is found in a gradient fraction with a higher sucrose concentration (more dense solution).

16c. The plasma membrane represents the least dense fraction because it has been fragmented into small pieces that re-seal to form small vesicle-like structures.

For further study, see text section "Purification of Cells and Their Parts," p. 152, and "Organelles of the Eukaryotic Cell," p. 168.

17a. The disappearance of the sucrosomes suggests that the added invertase is taken up by the cells and becomes localized in lysosomes, where it catalyzes breakdown of the sucrose. Additional experiments to test this hypothesis might include labeling invertase with a fluorescent or radioactive marker, adding it to cells, isolating the lysosomes, and determining if the invertase activity co-purified with acid phosphatase or some other lysosomal marker molecule. In fact, if you prepared a fluorescent invertase preparation, you might even be able to detect lysosomal fluorescence using a fluorescence microscope.

17b. Invertase in the fused cells also is located in a lysosomal compartment; this conclusion is consistent with the hypothesis discussed above. The observations may or may not be consistent with other hypotheses that you might have formulated. Since the sucrosomes in the fused cells disappeared over time, the invertase in the lysosomes derived from invertase⁺ cells must come into contact with and break down the sucrose in the lysosomes derived from sucrose⁺ cells, indicating that lysosomes inside cells mix rapidly, probably by fusion of their membranes.

For further study, see text section "Lysosomes are Acidic Organelles That Contain a Battery of Degradative Enzymes," p. 169.

6

Manipulating Cells and Viruses in Culture

PART A: *Chapter Summary*

Molecular cell biologists use clones of bacteria, simple eukaryotes such as yeast, and mammalian cells to study genetics and metabolism. These clones, identical descendents of a single cell, can be grown in suspension culture or on a solid surface. Bacteria can be grown in simple or minimal media, which are ideal for isolating nutritional auxotrophs by replica plating. Animal cells require richer, more complex media for growth. Two kinds of animal cells are commonly used for research: primary cells derived from an animal that have a limited lifespan in culture, or transformed cells, often obtained from tumors, that grow indefinitely. Animal cells can be fused to identify the chromosomal location of genes. A Nobel Prize–winning technique uses cell fusion to obtain cells that grow forever and produce monoclonal antibodies that react with a unique epitope on a protein.

Viruses are simple biological entities that reproduce only within living cells. They have a small number of genes and are excellent model systems for studying cellular processes. Viruses consist minimally of one type of nucleic acid surrounded by a protein coat called a capsid. Some viruses have an additional layer, the phospholipid envelope, derived from the host cell. The steps in the lytic viral life cycle are adsorption, penetration, replication (production of progeny nucleic acid and of viral proteins on cellular ribosomes), and release. Some bacterial viruses alternate between lysis and lysogeny, during which the viral genome integrates into the bacterial chromosome. The same phenomenon occurs with animal retroviruses such as HIV. Integrating viruses are currently being engineered as gene therapy vectors to transduce normal genes into organisms with genetic defects.

PART B: *Reviewing Concepts*

6.1 *Growth of Microorganisms in Culture*
(lecture date _____)

1. How many cells would you need to assay to isolate a mutation in any gene in *Escherichia coli*?

6.2 *Growth of Animal Cells in Culture*
(lecture date _____)

2. Although the amino acid glutamine is a nonessential amino acid, cultured human fibroblasts grow only in the presence of this amino acid. Explain this observation.

3. What are the advantages and disadvantages of using cultured cells rather than a whole organ, such as the liver, in molecular cell biology research?

4. What are the differences between monoclonal and polyclonal antibodies?

5. State one disadvantage of using transformed cells to study virus replication.

6.3 Viruses: Structure, Function, and Uses
(lecture date _____)

6. If you had a cell-free extract that was capable of in vitro protein synthesis, how would you use this extract to determine if the single-stranded RNA genome of a newly discovered virus were of plus or minus polarity?

7. Some viruses have an outer phospholipid envelope, which is derived from the plasma membrane of the host cell but contains virus-coded glycoproteins. Experimentally, such enveloped viruses are used to promote fusion of animal cells. What is the role of these viral glycoproteins in the viral infectious cycle?

8. If you compared the tertiary structure of the virion proteins of icosahedral viruses infecting animals and plants, what would you expect to find and why?

9. What characteristics of viruses make them good choices for gene therapy agents?.

10. What characteristics of retroviruses make them less than ideal candidates for gene therapy agents?

Clone no.	Thymidine Kinase *	Human chromosomes
1	+	9, 11, 17, 20
2	+	3, 11, 17, 19
3	-	2, 3, 11, 20
4	+	16, 17, 21
5	+	9, 12, 17, 19
6	-	9, 16, 21
7	+	2, 17

* (+) indicates presence and (–) indicates absence of thymidine kinase.

Table 6-1

Part C: Analyzing Experiments

6.2 Growth of Animal Cells in Culture

11. (••) A mouse cell lacking the enzyme thymidine kinase was fused with a primary human cell containing this enzyme. The cells then were grown in the absence of thymidine. After several generations of growth, several cell clones were isolated and assayed for thymidine kinase. In addition, the human chromosomes retained by these hybrids were identified based on their banding patterns.

   a. Based on the resulting data shown in Table 6-1, what can you conclude about the chromosomal location of the gene encoding thymidine kinase?

   b. How could you confirm your conclusion experimentally?

12. (•••) You have just been hired by a company that produces commercial monoclonal antibodies. Your immediate assignment is to establish pilot-scale conditions for production of a new monoclonal antibody, termed 9E11, to be used in a home diagnostic kit for prostate cancer. Until now, the 9E11 hybridoma has been grown in a medium supplemented with fetal calf serum. Your boss asks you whether switching to a defined medium would be desirable.

   a. What are the differences between a defined medium and a serum-supplemented medium?

   b. Might the decision be different if the 9E11 antibody was going to be marketed as an immunotherapeutic drug rather than as a immunodiagnostic reagent?

Amount of viral solution added	Plaques/dish (3 replicates)
UNDILUTED SOLUTION	
1.0 ml	No cells remaining in any culture
0.1 ml	No cells remaining in any culture
DILUTION 1 (1 part virus and 999 parts buffer)	
1.0 ml	No cells remaining in any culture
0.1 ml	Too many plaques to count
DILUTION 2 (1 part virus and 99,999 parts buffer)	
1.0 ml	343, 381, 364
0.1 ml	33, 38, 41

Table 6-2

6.3 Viruses: Structure, Function, and Uses

13. (••) You wish to assay a solution of Newcastle disease virus (NDV) sent to you by another investigator. One way to quantitate this virus is to measure the activity of a virally coded enzyme called hemagglutinin, which is present on both infectious and noninfectious viral particles. Previous investigators have found that there are 1×10^6 NDV particles per hemagglutinin unit (HU). A hemagglutinin assay of your NDV solution indicates that it contains 400 hemagglutinin units per ml. Next you perform a plaque assay by inoculating petri dishes containing a complete sheet (confluent monolayer) of chicken fibroblasts with various amounts of the NDV solution; 3 days later you count the cleared areas in the cell monolayer. These data are shown in Table 6-2.

   a. What is the total concentration of NDV particles in the original solution?

   b. What is the concentration of infectious viral particles in the original solution?

   c. What percentage of the NDV particles in the original solution are infectious?

14. (•••) David Baltimore and Howard Temin independently discovered the enzyme reverse transcriptase in retroviruses. (Their experiments are described in Classic Experiment 6.1 on the MCB 4.0 CD-ROM.) This enzyme uses RNA as a template for the synthesis of DNA.

   a. Why did this discovery go against the "Central Dogma?"

   b. How did Baltimore and Temin demonstrate that DNA was being synthesized by the enzyme?

   c. How did Baltimore and Temin demonstrate that the template for synthesis was RNA?

   d. Baltimore did an experiment in which he used both dNTP and NTPs as precursors for nucleic acid synthesis with a retrovirus and vesicular stomatitis virus, (another virus with an RNA genome that is not a retrovirus) as template. The results showed that the retrovirus polymerized dNTPs but not NTPs, while VSV polymerized NTPs but not dNTPs. Why did he do this experiment?

Answers

1. The genome of *Escherichia coli* is 4×10^6 base pairs (bp). In one generation, the probability of a single spontaneous mutation is about 1 in $10^6 - 10^8$ bp. If one mutation occurs in 10^6 bp, there could be four mutations in one genome. In order to detect one, it would be necessary to plate $10^6 \div 4$ or 2.5×10^5 cells. If there were one mutation in 10^8 bp, then $10^8 \div 4$ or 2.5×10^7 cells would be required. In reality, some base changes do not cause amino acid substitutions and not all amino acid substitutions cause a loss of protein function. Therefore, the actual number of bacterial cells required would be larger.

   For further study, see text section "Growth of Microorganisms in Culture," p. 182.

2. Sufficient quantities of glutamine to supply the entire organism are synthesized in human liver and kidney cells. It is therefore classified as a nonessential amino acid. However, because this source of glutamine is not available to cultured fibroblasts, which do not synthesize glutamine, it must be supplied in the medium.

For further study, see text section "Growth of Animal Cells in Culture," p. 183.

3. Cultured cells are preferable to whole organs because they consist of a single cell type and can be derived from organisms that are not routinely used as experimental animals (e.g., humans). In addition, environmental and genetic variables can be more closely monitored (if not controlled) with cultured cells than with organs. A disadvantage of cultured cells is that cell-cell interactions, which are present in an organ and which may be important determinants of the process under study, are abnormal in cell cultures. Also, biosynthesis—especially of tissue-specific macromolecules (e.g., glutamine synthetase in the liver)—may be low or nonexistent in cultured cells and very high in the intact organ.

For further study, see text section "Growth of Animal Cells in Culture," p. 181.

4. Monoclonal antibodies react with only one antigenic determinant (epitope) on a protein molecules, even though multiple epitopes are present. Polyclonal antibody preparations are a mixture of monoclonal antibodies.

For further study, see text section "Growth of Animal Cells in Culture," pp 189–190.

5. Most transformed cells are aneuploid, lacking some chromosomes found in their non-cancerous counterparts. The missing chromosomes could carry genes that encode cellular proteins required for viral replication.

For further study, see text section "Growth of Animal Cells in Culture," p. 187.

6. By convention, the polarity of mRNA is designated as plus. Thus, if the RNA of the virus were of plus polarity, it would act as mRNA when added to the extract and would direct the incorporation of a radioactive amino acid into a polypeptide, which could be precipitated by trichloroacetic acid.

For further study, see text section "Viruses: Structure, Function, and Uses," Figure 6–20, p. 199.

7. The viral proteins in the envelope interact with a specific membrane protein (the viral receptor) on the surface of a host cell, thereby promoting binding, or adsorption, of the virion to the host cell. This is the first step in the infectious cycle.

For further study, see text section "Viruses: Structure, Function, and Uses," p. 196.

8. The tertiary structure of the virion proteins would contain similar elements, because there are only a limited number of ways in which these proteins can be folded in order to maintain the overall icosahedral shape of these viruses. Differences do exist among the proteins of even closely related icosahedral viruses; these differences are often located on the surface of the virion and define the viral antigenic sites and the points at which the virus interacts with the cell.

For further study, see text section "Viruses: Structure, Function, and Uses," pp. 192–193.

9. Viruses are good choices for gene therapy vectors for several reasons. Their genomes are small so the recombinant DNA manipulations required to insert the normal gene into the viral genome are relatively easy. Since some viruses infect only a limited number of tissues, it may be possible to select a particular virus that will infect the tissue in which the wild-type gene must be expressed in order to overcome the disease. Some viruses integrate, providing for longer-term expression of the transgene.

For further study, see text section "Viruses: Structure, Function, and Uses," p. 203.

10. Retroviruses integrate randomly in the cell genome and may therefore disrupt a necessary cellular function. Also, retroviruses contain powerful initiators of transcription and may cause overproduction of cellular proteins, with unfortunate consequences.

For further study, see text section "Viruses: Structure, Function, and Uses," p. 203.

11a. The gene encoding thymidine kinase (TK) is on human chromosome 17, the only one present in all TK⁺ clones and absent from TK⁻ clones. This type of analysis, called *concordance analysis,* correlates the presence of retained human chromosomes with the presence of a particular trait and takes advantage of the fact that the hybrid cells lose human chromosomes in a seemingly random manner.

11b. In order to confirm this conclusion, one could test whether the observed TK activity is due to the human enzyme by use of specific antibodies, electrophoretic analysis, kinetic characteristics, etc. Such experiments could eliminate the possibility that the observed activity is due to some other enzyme or to reversion of the mouse TK⁻ mutation, rather than to the presence of human TK.

For further study, see text section "Growth of Animal Cells in Culture," p. 187.

12a. Defined, or basal, medium consists entirely of substances (salts, vitamins, amino acids, carbon source, etc.) that are present in known amounts. Although most bacterial cells will grow in relatively simple defined media, the most common media for culturing animal cells are supplemented with animal serum as a source of growth factors. In addition to growth factors, serum contains numerous other proteins. The exact composition of the serum, and hence of a supplemented medium, varies with the animal source and the environmental conditions under which it lives. See MCB, Table 6-3 (p. 194).

12b. If a human monoclonal antibody is to be injected into patients as an immunotherapeutic agent, then the purity of the antibody becomes very important. Since contaminating fetal calf serum proteins might induce an immune reaction, use of a defined medium that supports antibody production probably would be preferred in this case.

For further study, see text section "Growth of Animal Cells in Culture," pp. 183-184.

13a. The total concentration of viral particles is calculated from the hemagglutinin assay results: (400 HU/ml) × (1 × 10⁶ particles/HU) = 4 × 10⁸ particles/ml.

13b. The concentration of infectious viral particles is calculated from the results of the plaque assay. Averaging the number of plaques that formed with dilution 2 and adjusting for the dilution factor gives the following:

$$\frac{368 \text{ plaques/ml diluted soln}}{1 \times 10^{-6} \text{ ml orig. soln/ml diluted soln}}$$

$$= 3.68 \times 10^8 \text{ infectious particles/ml}$$

13c. 92 percent of the viral particles are infectious, as follows:

$$3.68 \times 10^8 \text{ infectious particles/ml} \times 0.92$$

$$= 4 \times 10^8 \text{ particles/ml}$$

For further study, see text section "Viruses: Structure, Function, and Uses," p. 194.

14a. The "Central Dogma" states the information flows unidirectionally and irreversible from DNA to RNA to protein. The discovery of reverse transcriptase showed that the first step was reversible. To date, the second step has not be challenged.

14b. Synthesis of DNA was shown by the DNase sensitivity of the product and the incorporation of radioactive dNTPs into a polynucleotide. This synthesis was inhibited by the substitution of an NTP for one of dNTPs.

14c. They treated the virus with RNase before attempting nucleic acid synthesis with dNTP. This pre-treatement prevented synthesis.

14d. This experiment showed that there was no RNA polymerase activity present in retroviruses that could also polymerize dNTPs.

7

Recombinant DNA Technology

PART A: *Chapter Summary*

The methods of recombinant DNA technology have revolutionized biology. Scientists now have the ability to produce large amounts of virtually any nucleic acid. Restriction endonucleases recognize a 4–6 nucleotide sequence in any double-stranded DNA and cut that DNA at a specific site within the sequence. The resulting pieces of DNA can be cloned by insertion into other biological molecules such as plasmids or bacteriophages that have been digested with the same enzyme; such vectors replicate to high levels in bacteria, and the amplified insert can be analyzed in more detail. Its size can be determined by electrophoresis on agarose or acrylamide gels; internal restriction endonuclease recognition sites can be located; and the absolute nucleotide sequence of the fragment can be ascertained by Maxam-Gilbert or Sanger dideoxy sequencing.

The same cloning techniques can produce libraries—collections of large numbers of bacteria harboring plasmids or phage, each of which contains a unique DNA. Genomic libraries contain large inserts of chromosomal DNA while cDNA libraries contain dsDNA copies of expressed mRNA. Libraries allow scientists to start with a gene and identify the protein it encodes, or to start with a purified protein and isolate its gene. The volume of a library that contains the gene for a particular protein is identified in a mutated organism. This clone is used to isolate the cDNA for the protein, which is then sequenced and expressed from a specialized vector. Alternatively, an amino acid sequence is obtained by biochemical means from a purified protein. An oligonucleotide whose sequence corresponds to the amino acid coding sequence is synthesized by chemical means. This oligonucleotide is used as a probe to screen a cDNA or genomic library. Alternatively, antibody can be raised to the purified protein, which can be used to screen a cDNA expression library. The positive clone can be used as a probe against a genomic library to isolate the gene.

The polymerase chain reaction (PCR) has resulted in a second revolution in molecular biology. Large amounts of nucleic acid can be produced starting from minute amounts of template if a small amount of sequence at the ends is known. This sequence is used to design primers, which initiate repeated doublings of the sequence between them in an exponential fashion.

Once the nucleotide sequence is available, it is stored, distributed, and analyzed with computers, using the methods of bioinformatics. Clues about function can be obtained by comparison with sequences of other genes/proteins contained in data banks. The complete nucleotide sequence of model organisms is either complete, or will be completed within a few years. The analysis and comparison of these sequences is called genomics. The nature of genes in the three major cell lineages, the Arachaea, the bacteria, and the eukaryotes, has been compared, demonstrating commonalities as well as differences reflecting their intra- and extracellular environments. The combination of complete genome sequences and microarray technology promises the elucidation of the coordination of gene expression in a variety of organisms.

PART B: *Reviewing Concepts*

7.1 *DNA Cloning with Plasmid Vectors*
(lecture date _____)

1. Plasmid cloning permits separation of different DNA fragments in a complex mixture. What are the two basic steps in this procedure? How is the separation of DNA fragments achieved?

7.2 *Constructing DNA Libraries with λ Phage and Other Cloning Vectors*
(lecture date _____)

2. What is the advantage of using λ phage vectors rather than plasmid vectors for producing and screening a genomic library, particularly for higher organisms?

3. You wish to find the amino acid sequence of a protein and have available positive clones from both cDNA and genomic libraries. Which would you use to find the sequence and why?

7.3 *Identifying, Analyzing, and Sequencing Cloned DNA* (lecture date _____)

4. In the Sanger method for sequencing DNA, four reaction mixtures are prepared, each containing DNA polymerase, all four normal deoxyribonucleoside triphosphates (dNTPs), and one of the four corresponding dideoxyribonucleoside triphosphates (ddNTPs). Why are both dNTPs and ddNTPs used in these reaction mixtures?

5. What is the advantage of using an EST rather than protein sequence to design a probe to screen a library?

7.4 *Bioinformatics* (lecture date _____)

6. How is bioinformatics used to find the function of a gene responsible for an inherited disease?

7.5 *Analyzing Specific Nucleic Acids in Complex Mixtures* (lecture date _____)

7. A first-year graduate student is running his first Southern blot. He's alone in the lab at 10 p.m. and can't find the alkaline buffer for transferring DNA from the gel to the membrane, so he uses 0.01 M Tris buffer, pH 7.5. Why is his blot unsuccessful?

7.6 *Producing High Levels of Proteins from Cloned cDNAs* (lecture date _____)

8. Some eukaryotic proteins can be expressed in bacteria, but some must be expressed in eukaryotic cells. What determines which expression system is used?

7.7 *Polymerase Chain Reaction: An Alternative to Cloning* (lecture date _____)

9. Why do many investigators, in selecting a DNA sequence to be used as a template for construction of PCR primers, look for sequences rich in codons for tryptophan or methionine?

10. What property of the DNA polymerase used in the polymerase chain reaction allows this technique to be successful?

7.8 *DNA Microarrays: Analyzing Genome-wide Expression* (lecture date _____)

11. Why would microchip assays to study gene expression be impossible without the science of Genomics?

PART C: *Analyzing Experiments*

7.3 *Identifying, Analyzing, and Sequencing Cloned DNA*

12. (••) You have cloned a cDNA that has the restriction map given below (Figure 7-1). (The distances between sites are give in kbp.) To create a mutation in the gene, you wish to reverse the orientation of the HindIII fragment by standard cloning procedures. How would you detect the clones that contained the desired plasmid?

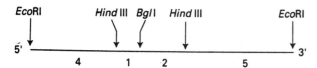

Figure 7-1

13. (••) You have discovered a virus with a circular double-stranded DNA genome containing approximately 10,000 bp. You want to begin characterizing this genome by making a map of the cleavage sites of three restriction endonucleases: EcoRI, HindIII, and BamHI. You digest the viral DNA under conditions that allow the endonuclease reactions to go to completion and then subject the digested DNA to electrophoresis on agarose to determine the lengths of the restriction fragments produced in each reaction. Based on the resulting data, shown in Table 7-1, draw a map of the viral chromosome indicating the relative positions of the cleavage sites for these restriction endonucleases.

Endonuclease	Length of fragments (kb)
EcoRI	6.9, 3.1
HindIII	5.1, 4.4, 0.5
BamHI	10.0
EcoRI + HindIII	3.6, 3.3, 1.5, 1.1, 0.5
EcoRI + BamHI	5.1, 3.1, 1.8
HindIII + BamHI	4.4, 3.3, 1.8, 0.5
EcoRI + HindIII + BamHI	3.3, 1.8, 1.5, 1.1, 0.5

Table 7-1

7.2 *Constructing DNA Libraries with λ Phage and Other Cloning Vectors*

14. (•••) Explain why vertebrate genomic libraries are often made from embryonic or sperm DNA.

15. (•••) Analysis of the DNA content of *Drosophila melanogaster* has shown that a haploid cell contains about 1.5×10^8 bp.

   a. How many standard λ phage vectors carrying 20-kb DNA fragments theoretically are required to constitute a complete *D. melanogaster* genomic library? How many vectors should you prepare in order to ensure that every sequence is included in the library?

   b. DNA fragments as long as 300 kb can be cloned in artificial yeast chromosomes (YACs). How many yeast clones theoretically are needed to contain all the genes of *D. melanogaster* using this cloning system? How many yeast clones should you prepare in order to ensure that every sequence is included in the library?

7.3 *Identifying, Analyzing, and Sequencing Cloned DNA*

16. (•) You want to clone the gene encoding a particular protein (P) in order to study the characteristics of the gene. You have an antibody to protein P, a cell line that expresses P at a reasonable level, and a genomic library of this cell line. Standard molecular cell biology equipment and technology also are available. Describe two general approaches you could use to isolate a genomic clone containing the gene encoding protein P.

17. (•••) You are the instructor for a laboratory course and have just lead the class through sequencing of a DNA fragment with the Sanger dideoxy method. Now it is time to help the students interpret the autoradiograms from their sequencing gels. The four gel lanes—A, C, G, T—obtained by Bob, Amy, and Ngai are shown in Figure 7-2. These students ask you two questions, as follows. How would you answer them?

   a. Why does the A sequence lane contain few bands at the bottom of the gel and a large "blob" of radioactivity at the top of the gel?

   b. Why does the intensity of the bands in a sequencing gel (e.g., lane C) vary?

A C G T

Figure 7-2

7.5 *Analyzing Specific Nucleic Acids in Complex Mixtures*

18. (••) You have available the genome of a eukaryotic dsDNA virus and cloned restriction fragments representing all portions of this genome. A map of the genome is shown in Figure 7-3. with the distances between sites given in kbp.

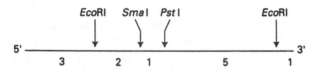

EcoRI   SmaI   PstI   EcoRI

5'——————————————————3'

3   2   1   5   1

Figure 7-3

a. How would you determine the sizes of the mRNAs encoded by the genome?

b. You believe that a particular gene is located within the 3 kbp Eco-Pst fragment. When you use this fragment as probe, a positive reaction is seen with an mRNA of 2.5 kbp. How would you explain this observation?

c. What experiment would you do to confirm this hypothesis?

Answers

1. The first step in plasmid cloning is to mix plasmids with a mixture of DNA fragments in the presence of DNA ligase. In this ligation step, each fragment is inserted into a plasmid molecule, producing a mixture of recombinant plasmids. In the second step, E. coli cells are incubated with the recombinant plasmids under conditions that promote transformation. Because each cell picks up only one recombinant plasmid, separation of the DNA fragments occurs during this step. Following plating of the transformed cells, each colony that grows will contain a single type of recombinant plasmid, which can then be isolated.

   For further study, see text section "DNA Cloning with Plasmid Vectors," pp. 212–214.

2. The main advantage of λ phage cloning vectors over plasmid vectors comes at the screening step, since many more phage plaques than bacterial colonies can be accommodated per area of a plate. Using phage vectors decreases the number of repetitive and costly manipulations to identify a desired recombinant.

   For further study, see text section "Constructing DNA Libraries with λ Phage and Other Cloning Vectors," p. 218

3. You would use the cDNA since the inserts in such libraries are smaller than those in genomic libraries, reducing the amount of sequencing required. As well, the cDNA library lacks introns, so that the sequence obtained should be that of the exon for the proteins; you would not have to distinguish which sequence is from an ORF and which from an intron.

   For further study, see text section "Constructing DNA Libraries with λ-Phage and Other Cloning Vectors," pp. 219–221.

4. In the Sanger method, the DNA to be sequenced is used as a template for synthesis of a complementary strand from a short deoxynucleotide primer. Although ddNTPs can be added to a nascent chain in place of the corresponding dNTPs, addition of a ddNTP prevents further chain elongation. In the presence of multiple copies of the

template, all four dNTPs, and one of the ddNTPs, DNA polymerase incorporates the dNTPs until a ddNTP molecule is incorporated, thereby terminating chain elongation. On different templates, incorporation of ddNTP and hence chain termination occurs randomly in the sequence of the growing chains. For example, in a reaction with dGTP and ddGTP, each time a C is present in the template, the polymerase "selects" either the deoxy or dideoxy form for incorporation into the nascent chain. In this fashion, some chains will be terminated at a specific G, while other chains will be extended past this G. This same decision process will occur each time there is a C in the template strand, resulting in a nested set of products with some ending at each and every G.

For further study, see text section "Identifying, Analyzing, and Sequencing Cloned DNA," p. 232.

5. The EST is derived from the mRNA sequence used for translation. This sequence reflects the codons used for translation without wobble.

For further study, see text section "Identifying, Analyzing, and Sequencing Cloned DNA," p. 227.

6. Once the gene for the mutated protein is cloned and sequenced, the sequence of the gene is compared to the sequences of all genes that are stored in data bases. Matches are found reflecting the conservation of amino acid sequence in functional domains in proteins. Even if the protein that matches is from a distantly related organism, the role that the two proteins play in cell metabolism may be similar; and loss of this function may correlate with the biological manifestation of the disease.

For further study, see text section "Bioinformatics," p. 235.

7. The blot is unsuccessful because the alkaline solution used for transfer denatures the DNA strands. If the DNA attached to the membrane is not single-stranded, the single-stranded probe will not hybridize.

For further study, see text section "Identifying, Analyzing, and Sequencing Cloned DNA," p. 240.

8. Proteins that are post-translationally modified by glycosylation or phosphorylation must be produced in eukaryotic cells, since bacterial cells do not carry out these biological functions.

For further study, see text section "Producing High Levels of Proteins from Cloned cDNAs," p. 245.

9. Hybridization of oligonucleotides is best if all bases hybridize to the template. Since tryptophan and methionine have only one codon each (i.e., there is no "wobble" in the third position), primers containing these codons have a better chance of hybridizing to a unique site in the template. If primers must be used that contain other codons, hybridization can be optimized by knowing the codon bias for the organism, so that the most prevalent codon for a particular amino acid can be selected, or by preparing a degenerate primer, with multiple bases for a codon.

For further study, see text section "Polymerase Chain Reaction: An Alternative to Cloning," p. 246. See also p. 226.

10. The polymerase used for PCR is derived from *Thermus aquaticus*. It maintains its activity at 72°C and is not denatured at the temperatures required for strand separation for reannealing of primer.

For further study, see text section "Polymerase Chain Reaction: An Alternative to Cloning," p. 246.

11. Microarray technology allows the analysis of gene expression by hybridization. In order for this strategy to be successful, the targets are the entire repertoire of nuclear genes. Without genomics, the determination of the entire nucleotide sequence of an organism, such targets would not be available.

For further study, see text section "DNA Microarrays: Analyzing Genome-wide Expression," pp. 238–249.

12. You would digest the resulting plasmids with Eco R1 and BglI simultaneously and determine the size of the inserts by agarose gel electrophoresis. If the original orientation were maintained, the fragments would be 5 and 7 kbp; if the desired orientation were obtained, there would be one band at 6 kbp that would be intense.

For further study, see text section "Identifying, Analyzing, and Sequencing Cloned DNA," pp. 230–231.

13. The map is given in Figure 7-4.

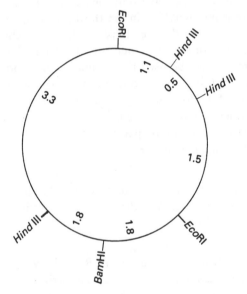

Figure 7-4

For further study, see text section "Identifying, Analyzing, and Sequencing Cloned DNA," pp. 230–231.

14. Presumably both embryonic and sperm DNA contain all the DNA sequences found in an organism. If differentiation is accompanied by loss of specific DNA sequences, which, for example, is known to occur in white blood cells, a genomic library prepared from adult organs or cells might be incomplete.

For further study, see text section "Constructing DNA Libraries with λ Phage and Other Cloning Vectors," p. 218.

15a. About 7500 clones ($1.5 \times 10^8 \div 2 \times 10^4$) theoretically are needed to constitute the Drosophila genome. However, statistical considerations (Poisson distribution) indicate that in order to ensure that every sequence has a 95 percent chance of being represented at least once, you should prepare a library containing about five times the theoretical minimum. The actual library, then, would contain 3.75×10^4 clones for the Drosophila genome.

15b. About 500 yeast clones ($1.5 \times 10^8 \div 3 \times 10^5$) theoretically could represent the entire Drosophila genome, and 2500 clones would form a useful library.

For further study, see text section "Constructing DNA Libraries with λ Phage and Other Cloning Vectors," p. 219.

16. To isolate a specific gene from a genomic library, you first could prepare either a cDNA probe or synthetic oligonucleotide probe that hybridizes to the clone containing the gene. In the example here, you could obtain a cDNA probe by using the antibody to precipitate polyribosomes containing nascent chains of protein P from the cell line that expresses it. The mRNA species still attached to the polyribosomes should be greatly enriched for the mRNA encoding this protein, since the antibody should only recognize (and precipitate) polyribosomes containing nascent chains of P. The cDNA corresponding to this mRNA then is prepared with reverse transcriptase in the presence of radiolabeled dNTPs. To obtain an oligonucleotide probe, you could use the antibody to isolate protein P by immunoaffinity chromatography and then sequence small portions of protein P by automated Edman degradation. These short amino acid sequences then are analyzed to determine the shortest, least degenerate oligonucleotide sequences that could encode the sequenced peptides. The necessary oligonucleotides then are synthesized chemically and end-labeled with [γ-^{32}P]ATP using polynucleotide kinase. Once you have prepared a suitable cDNA or oligonucleotide probe, you can then use it in a membrane-hybridization assay to screen a genomic library from the animal or plant used to make the cell line. Any genomic clones that hybridize to the probe can then be investigated for the presence of introns, transcription start sites, and other attributes of an active gene as discussed in later chapters.

For further study, see text section "Identifying, Analyzing, and Sequencing Cloned DNA," pp. 223–227.

17a. In the Sanger dideoxy method, incorporation of a dideoxynucleotide results in chain termination. If the concentration of a specific dideoxynucleoside triphosphate is too low relative to its normal counterpart, then the frequency of chain termination will be low and relatively few short, truncated chains will be formed. As a result, few short chains will be visualized at the bottom of the sequencing gel and a "blob" of poorly resolved longer DNA chains will be visualized at the top of the gel.

17b. The variation in band intensity in a Sanger sequencing gel results from the variable rate at which DNA polymerase replicates DNA. In relative terms, the polymerase pauses at certain positions, so that growing DNA molecules ending at these positions accumulate. If a dideoxynucleotide is added at these "pause" positions, thereby causing termination, the corresponding truncated molecules will be over represented in the population resolved on the gel, causing a darker band.

For further study, see text section "Identifying, Analyzing, and Sequencing Cloned DNA," pp. 232–233.

18a. You would carry out a Northern blot of mRNA isolated from infected cells using the viral genome as probe.

18b. The gene contains an intron.

18c. You would hybridize the radioactive EcoR1–Sma I fragment and the radioactive Sma I–PstI fragments to mRNA, digest with S1 nuclease, electrophorese the digest on an agarose gel, and expose the dried gel to x-ray film. The resulting nucleic acids should be smaller than the restriction fragments. This gives a preliminary indication of the location of the intron on the genome. Its exact location can be obtained by comparing the genome sequence to the sequence of cDNAs of viral mRNAs.

For further study, see text section "Analyzing Specific Nucleic Acids in Complex Mixtures," pp. 241–242.

8 Genetic Analysis in Cell Biology

PART A: *Chapter Summary*

Classical genetics in combination with recombinant DNA technology is a powerful tool to analyze the function of proteins in the cell and the organism. The function of the protein may be inferred by the consequences of the loss of its gene, the effect of mutation in the genotype on the phenotype. Mutations occur spontaneously or can be induced by various means, such as the application of chemicals or recombinant DNA methods. Mutations can be as small as a single base change or so massive that deletion or rearrangement of portions of chromosomes can be seen through the light microscope.

Predicting the phenotypes of offspring carrying different types of mutant alleles (recessive or dominant) is possible because of the regular assortment of chromosomes during meiosis in diploid organisms. In haploid organisms, all mutations are essentially dominant. Specialized genetic screens can show the number of genes involved in a biological process through complementation analysis, the sequence in which they act in a metabolic pathway, and whether two genes interact *in vivo* for activity.

The program of contemporary genetics is to start with a mutant gene and identify the protein coded by its wild-type allele. Through positional cloning, a mutation-defined gene is isolated by correlating the genetic map with a physical map of the chromosome. A genetic map shows the position of genes relative to each other; they can be on the same chromosome, on different chromosomes, or on the sex chromosome. If they are on the same chromosome, the distance between them is calculated by recombinational analysis. In contrast, a physical map is the nucleotide sequence of a chromosome. Such a map exists only for a few model organisms; for other organisms, available DNA clones can be ordered to reflect their position on the chromosome. Positional cloning begins by identifying a cloned DNA fragment near the mutated gene by *in situ* hybridization, chromosome abnormalities, or the presence of markers, such as restriction fragment length polymorphisms, linked to the mutated gene. This starting clone initiates a chromosome walk, moving closer to the clone containg the gene by isolating overlapping DNA fragments. Once a candidate clone is in hand, the site of the gene within the clone is found by a variety of methods including recombinational analysis with linked markers, including RFLPs, Southern blotting, identification of the mRNA transcribed in proximity to the gene, detection of single base changes, and sequencing. Once the sequence is known and converted to amino acid sequence, comparison with other sequences stored in computer banks allows the function of the protein to be deduced.

Another powerful technique to study the effect of mutation on cells and organisms is to introduce a gene with a specific mutation created *in vitro*. Through homologous recombination, the mutant gene replaces the wild-type gene, and the result of the loss can be observed. Such gene-targeted knockouts can be obtained in the germ line of mice and in yeast; useful mouse models of human cystic fibrosis have been achieved in this way. Transgenic organisms that have foreign genes inserted by nonhomologous recombination are used to identify a clone corresponding to a mutation-defined gene, produce models for diseases caused by dominant mutations, and, in the case of plants, create "factories" for the production of human proteins as pharmaceuticals.

PART B: *Reviewing Concepts*

8.1 *Mutations: Types and Causes*
(lecture date _____)

1. How would you determine if an organism that appeared to have a normal phenotype were homozygous or heterozygous for the wild-type allele?

2. What is a frame shift mutation? What other type of mutation can result from a frame shift mutation?

8.2 *Isolation and Analysis of Mutants*
(lecture date _____)

3. You have several virus strains carrying conditional mutations.

   a. What kinds of experiments could you perform with these mutants to study the nature and function of viral genes?

   b. Why can this type of analysis be carried out with viruses?

4. The genetic dissection of the arginine biosynthetic pathway in *N. crassa,* which was carried out by Beadle and Tatum, is depicted in Figure 8-1. Explain how this analysis led to the "one gene-one enzyme" postulate.

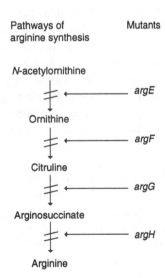

Figure 8-1

5. You find that a temperature-sensitive yeast mutation may be corrected by two different genes. How can two different genes correct the same temperature-sensitive mutation?

8.3 *Genetic Mapping of Mutations*
(lecture date _____)

6. Explain why the value of a centimorgan is not a constant.

7. What are the similarities and differences between RFLPs and genes?

8. How is a DNA polymorphism related to a restriction fragment length polymorphism (RFLP)?

9. In order to detect a RFLP linked to a mutant gene causing an inherited disease, multiple restriction endonucleases and multiple probes are used. Why?

8.4 *Genetic Mapping of Mutation*
(lecture date _____)

10. What is a physical map of a chromosome or region of a chromosome?

11. How are RFLPs useful in positional cloning?

8.5 *Gene Replacement and Transgenic Animals* (lecture date _____)

12. In the production of knockout and transgenic organisms, the introduced DNA often contains one or more marker genes in addition to the gene of interest (transgene).

    a. What is the purpose of including marker genes in such gene-transfer experiments?

    b. Give examples of marker genes used in gene-transfer experiments with yeast, mice, and *Drosophila.*

Part C: *Analyzing Experiments*

8.1 *Mutations: Types and Causes*

13. (••)The single-base change that causes sickle-cell anemia destroys one of the three *Dde*I sites in a portion of the normal B-globin gene. Affected individuals produce an abnormal hemoglobin (HbS) rather than wild-type HbA. You have three radioactive DNA probes (A, B, C), corresponding to the regions of the B-globin gene indicated in Figure 8-2.

    a. Draw a sketch of the banding patterns expected in Southern blots of the *Dde*I digests of the normal B-globin gene and of the mutant sickle-cell B-globin gene visualized with each of the three probes.

    b. Explain your sketch.

Figure 8-2

8.2 *Isolation and Analysis of Mutants*

14. (•)Cell division cycle (*cdc*) mutants in yeast are blocked at specific points in the cell cycle, and most such mutants are temperature sensitive.

    a. What is the easiest way to enrich for temperature-sensitive *cdc* mutants in a yeast culture? Would this procedure yield only *cdc* mutants?

    b. How can *cdc* mutants be distinguished from other temperature-sensitive mutants?

15. (••)The yeast genome, which contains 1.4×10^7 base pairs, is both smaller and less complicated than the human genome. In contrast to the human genome, the entire yeast genome can be carried on a few thousand plasmids. You want to screen a yeast genomic library for the *SEC18* gene in order to characterize the gene at the DNA level and ultimately its encoded protein. *SEC18* is one of several genes required for proper maturation of secretory proteins in the endoplasmic reticulum and Golgi complex.

    a. You have available a temperature-sensitive *sec18* mutant yeast strain. Describe a screening protocol using genetic complementation for identifying a genomic clone containing the *SEC18* gene.

    b. Generally, plasmid cloning vectors contain a selectable marker such as the *LEU2* gene, which encodes an enzyme essential for leucine biosynthesis. Assume that the plasmids constituting your genomic library contain *LEU2* and that the *sec18* mutant strain also carries a defective *LEU2*. What advantage might there be to including a leucine selection step in your screening protocol?

8.3 *Genetic Mapping of Mutations*

16. (•)Pedigrees of families affected by Cystic fibrosis (CF), Huntington's disease (HD), and Duchenne muscular dystrophy (DMD) are shown in Figure 8-3.

(a) Cystic fibrosis

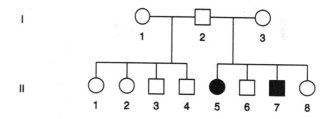

(b) Huntington's disease

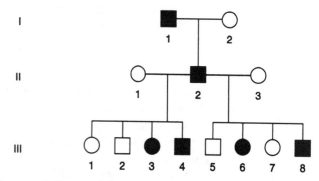

Figures 8-3 a and b

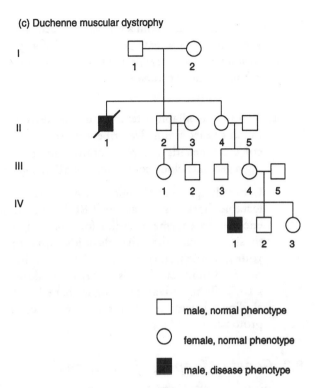

(c) Duchenne muscular dystrophy

□ male, normal phenotype

○ female, normal phenotype

■ male, disease phenotype

● female, disease phenotype

Figure 8-3 c

a. Based on these case studies, state whether each disorder is inherited autosomally or is sex linked and whether it is recessive or dominant. Explain your answers.

b. Phenotypically unaffected members of the human population may be carriers for a genetic disease. What are the odds that the children of parents, both of whom are carriers for an autosomal recessive defect, will be affected? What are the odds that the children of parents, one of whom is a heterozygous carrier of an autosomal dominant defect, will be affected? What are the odds that children of parents, one of whom is a carrier for an X-linked trait, will be affected?

c. How can a dominant lethal trait be maintained in the human gene pool?

17. (•)Assume that *a1* and *a2* are two alleles of the same gene; *b1* and *b2* are two alleles of another gene. A female mouse heterozygous for both genes mates with a male mouse homozygous for both genes, having only the *a2* and *b2* alleles. The offspring have two genotypes: *a1a2/b2b2* and *a2a2/b1b2*. Are genes *a* and *b* on the same or different chromosomes? Explain your answer.

8.4 *Molecular Cloning of Genes Defined by Mutations*

18. (•)DNA probes containing a reporter molecule can be used to map the distribution of genes along chromosomes. Compare the probable sensitivity of this type of *in situ* hybridization when the target chromosomes are polytene salivary gland chromosomes of *Drosophila melanogaster*, mitotic chromosomes from human skin fibroblasts, or fibers of chromosomes.

19. (••)Chromosome (DNA) walking is a molecular technique to isolate contiguous regions of genomic DNA beginning with a previously cloned DNA segment that maps near a gene of interest. DNA fragments from each contiguous region are used as probes in a reiterative, step-wise screening of a genomic library or libraries. Success in chromosome walking requires a genomic library containing fragments prepared by (i) incomplete digestion with one restriction enzyme, (ii) complete digestion with two distinct restriction enzymes, or (iii) physical fragmentation techniques such as shearing. Why is this so?

Answers

1. You would do a test cross, mating the organism with another that was known to be homozygous recessive. If the organism were homozygous dominant, all the progeny will appear dominant. If the individual were heterozygous for the trait, half of the offspring would appear dominant and half would appear recessive.

   For further study, see text section "Mutations: Types and Causes," p. 257, and Fig. 8-3.

2. A frame shift mutation is the addition or deletion of any number of nucleotides that is not a multiple of three. At some point past the additional or deletion, it is likely that a nonsense mutation will be created, resulting in inability to complete translation.

   For further study, see text section "Mutations: Types and Causes," p. 258 and Fig. 8-4.

3a. You could infect wild-type cells with mutant viruses and then shift the cells to the nonpermissive temperature at various times after infection. Analysis of the viral components synthesized before the shift could give an indication of the order in which processes occur during the infectious cycle.

3b. The analysis can be carried out because all viruses except for retroviruses are haploid.

For further study, see text section "Isolation and Analysis of Mutation," p. 265, for an analogous situation.

4. Beadle and Tatum's experiment was designed to determine the genetic regulation of a known biochemical pathway. Their rationale was to show that mutation blocked the formation of intermediates in the pathway. In their words, "...a mutant can be maintained and studied if it will grow on a medium to which has been added the essential product of the genetically blocked reaction." The link that lead to the "one gene-one enzyme" postulate was that if the gene were mutated, the reaction would not occur because the mutation caused an enzyme to be produced incorrectly or not at all. Some of Beadle and Tatum's early work was done with mutants in the synthesis of vitamins, whose biosynthesis was known to be catalyzed by enzymes. These mutants could be grown in minimal medium supplemented with the vitamin.

For further study, see text section "Isolation and Analysis of Mutation," p. 265 and Fig. 8-12.

5. Two different general mechanisms exist for correcting a genetic defect: complementation (i.e., introduction of the normal form of the mutated gene) and suppression. Suppression occurs when an alteration in one gene product corrects a defect in another gene product. For example, in the case of two proteins that interact with one another, a mutation in one protein that results in a defective protein-protein interaction may be suppressed by a compensating change in the second protein. The identification of such suppressor mutations in yeast has been a major approach to characterizing interacting proteins.

For further study, see text section "Isolation and Analysis of Mutation," pp. 265-266.

6. A centimorgan is defined as the distance separating two loci on a chromosome that exhibit a recombination frequency of 1 percent. Since the size of chromosomes differs, the distance required for genes to recombine at a 1 percent frequency is directly proportional to the length of the recombining chromosomes. Other factors characteristic of a particular pair of chromosomes also can influence the recombination frequency.

For further study, see text section "Genetic Mapping of Mutations," p. 270.

7. Like genes, RFLPs are heritable, can occur in the heterozygous state, and have an associated phenotype (the appearance of a fragment of DNA with a characteristic size after digestion of DNA with a particular restriction enzyme and Southern blotting). Unlike genes, RFLPs have no regulatory elements and encode no RNA or protein.

For further study, see text section "Genetic Mapping of Mutations," pp. 271-272.

8. A DNA polymorphism is simply a difference in DNA sequence between individuals at a particular site. When such changes occur in recognition sites for restriction endonucleases, the fragments resulting from digestion with the enzymes increase in size, if the change destroys the site, or decrease in size if a new site is created.

For further study, see text section "Genetic Mapping of Mutations," p. 271.

9. RFLPs arise because of mutations that either create or destroy the recognition sites for various restriction endonucleases. Such a mutation leads to a gain or loss of a particular recognition site, which is reflected by a change in the length of restriction fragments resulting from cleavage by the corresponding restriction enzyme. In searching for an RFLP in a specific DNA region, multiple restriction enzymes are used because the identity of the enzyme whose site has been affected is not known. Likewise, multiple probes initially must be

used in the Southern-blot analyses, since the sequence bounded by the restriction sites is not known. Consequently, the correct combination of endonuclease and probe must be found in order to demonstrate an RFLP. This can be a daunting task. In the identification of the RFLP associated with Huntington's disease, for example, the investigators considered themselves lucky when the 10th probe tested revealed a linked RFLP.

For further study, see text section "Genetic Mapping of Mutations," p. 271 and Fig. 8-20.

10. An ideal physical map is the complete nucleotide sequence of a chromosome. Until this data is available, a partial physical map can be a diagram that orders cloned fragments of DNA so that they reflect the order in which they would line up on a chromosome or fragment thereof. These fragments would contain the number of nucleotides between restriction endonuclease recognition sites within each fragment.

For further study, see text section "Genetic Mapping of Mutations," p. 266.

11. RFLPs are used twice in positional cloning. In the first step of the process, the availability of a large number of polymorphisms linked to the mutation defined gene aids in selecting a DNA clone as a starting point for a chromosome walk. Once a clone that is likely to contain the mutation-defined gene is identified, the position of the gene within this clone can in some cases be located by recombinational analysis between RFLPs, the gene in question, and another visible phenotypic marker.

For further study, see text section "Genetic Mapping of Mutations," pp. 272, 277, and Fig. 8-26.

12a. Marker genes are introduced along with the transgene in order to select for or identify recipients that carry the transgene.

12b. A marker gene used in production of knockout yeast is *URA3,* a mutant in uracil biosynthesis. Because the recipient cells require uracil, only cells that take up and integrate the linked transgene

can grow in the absence of the pyrimidine. See MCB, Figure 8-34 (p. 293). In the production of transgenic fruit flies, a marker gene that produces red eyes is located on the P element along with the transgene. The P element is injected into an embryo with white eyes. After transposition of the element to the chromosome, some flies will have both the transgene and the marker gene in their germ line. If these flies are mated with white-eyed flies, any progeny with red eyes contain the transgene. Several markers are used in the production of knockout mice. Two marker genes conferring resistance to two antibiotics—neomycin and ganciclovir—are introduced along with the disrupted transgene. Neomycin is used for positive selection of embryonic stem (ES) cells that have incorporated the transgene. Then ganciclovir is used to kill cells that did not insert the transgene in the correct site (negative selection). The ES cells also are homozygous for a marker gene conferring black coat color. If the ES cells are injected into embryos homozygous for the white-coat allele, only progeny mice with some black in their coats will carry the disrupted transgene of interest.

For further study, see text section "Gene Replacement and Transgenic Animals," pp. 282-289.

13. See Figure 8-4. Probe A corresponds to ≈90 nucleotides spanning the mutated *Dde*I site and includes sequences in both the 175-bp and 201-bp fragments resulting from *Dde*I digestion of the normal gene. Hence two bands should be seen in the Southern blot of the normal digest with probe A. Since the mutated *Dde*I site is missing in the sickle-cell DNA, digestion yields only one large fragment that is the sum of the two smaller fragments (376 bp). Probe A will reveal this single fragment. Probes B and C are shorter probes (≈20 nucleotides) located on either side of the *Dde*I site. Probe B includes sequences in the 175-bp fragment, and probe C includes sequences in the 201-bp fragment. Each probe will then reveal the corresponding single fragment of 175 or 201 bp when the normal digest is analyzed and one larger fragment of 376 bp with the sickle-cell digest.

For further study, see text section "Mutations: Types and Causes," pp. 258-259.

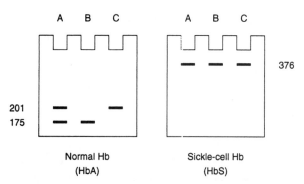

Figure 8-4

14a. Enrichment for temperature-sensitive (ts) *cdc* mutants is easily achieved by looking for clones that grow and divide at low temperatures (20–24°C) but not at elevated temperatures (37°C). However, most of the ts mutants obtained will be defective in some critical metabolic reaction, such as ATP production or DNA synthesis, and will not be true *cdc* mutants.

14b. True *cdc* mutants can be distinguished from nonspecific ts mutants because the *cdc* mutants will be arrested at a defined point in the cell cycle, whereas the nonspecific mutants will be arrested at random points in the cell cycle. A corollary of this observation is that *cdc* mutants should show synchronous entry into the cell cycle when shifted from the nonpermissive to the permissive temperature (commonly the lower temperature). If the mutant clone in question shows synchronous entry into the S phase (as measured by [³H]thymidine incorporation into DNA) when shifted from the high temperature to the low temperature, it is probably a *cdc* mutant.

For further study, see text section "Isolation and Analysis of Mutants," pp. 261-262 and Figure 8-9.

15a. By transfecting mutant cells with plasmids from the library, the library could be screened for its ability to correct (complement) the *sec18* mutation. In this approach, yeast cultures would be transfected with plasmids under conditions permissive for cell growth (i.e., at lower temperature) and maintained at this temperature for enough time to allow protein expression. The yeast cultures would then be plated out and incubated under nonpermissive conditions (i.e., at elevated temperature). Only cells that have been transfected with a plasmid encoding *SEC18* and thus capable of expressing wild-type Sec18 protein should grow at the nonpermissive temperature to generate a colony. The plasmid could then be isolated, the *SEC18* gene isolated, and its sequence determined. From this sequence, the amino acid sequence of the encoded protein can be deduced and compared with that of know proteins. Of course, such a screening protocol may, with low probability, isolate a suppressor gene rather than the SEC18 gene itself.

15b. The protocol described above is the simplest possible approach. It assumes that there is no need for replica plating to explicitly contrast growth at the permissive temperature versus the nonpermissive temperature and that transfected plasmids will be maintained in cells in the absence of any selective advantage. Because these assumptions often are not true, it would be advantageous to screen the plasmid library with a *sec18* mutant that also carries a *leu2* mutation. Since the plasmid vector contains the normal *LEU2* gene, mutant cells that have taken up a plasmid could be selected by first plating on leucine-deficient medium at the permissive temperature. The colonies that grow under these conditions then would be replica-plated at the nonpermissive temperature to identify cells carrying a plasmid containing the *SEC18* gene. An example of a standard protocol for genetic complementation cloning is shown in Figure 8-5.

For further study, see text section "Isolation and Analysis of Mutants," p. 265.

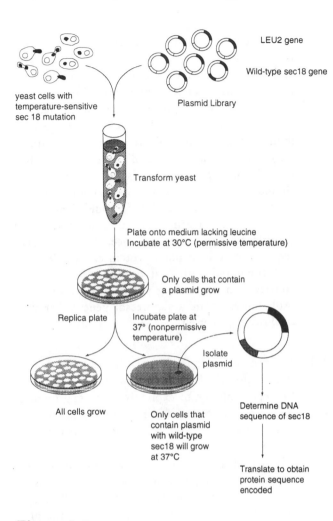

yeast cells with temperature-sensitive sec 18 mutation

LEU2 gene

Wild-type sec18 gene

Plasmid Library

Transform yeast

Plate onto medium lacking leucine
Incubate at 30°C (permissive temperature)

Only cells that contain a plasmid grow

Replica plate

Incubate plate at 37° (nonpermissive temperature)

Isolate plasmid

All cells grow

Only cells that contain plasmid with wild-type sec18 will grow at 37°C

Determine DNA sequence of sec18

Translate to obtain protein sequence encoded

Figure 8-5

16a. Cystic fibrosis must be an autosomal recessive disorder, since none of the parents in generation I show a phenotypic defect, but two out of the four children (both male and female) of parents I-2 and I-3 show a phenotypic defect. Parents I-2 and I-3 must be heterozygotes, whereas parent I-1 must be a homozygous normal since none of the four children of parents I-1 and I-2 show a phenotypic defect. Huntington's disease must be an autosomal dominant disorder, since at least one member of each generation shows a phenotypic defect and in generation III this defect is shown irrespective of sex. Duchenne muscular dystrophy must be an X-linked trait, which is not expressed on the Y chromosome of males, since all individuals showing the phenotypic defect are male, yet no parent in generations I-IV is phenotypically defective. The defective allele must be carried on the X chromosome of heterozygous females in which it

is recessive. For each generation, male children of a carrier female have a 50 percent probability of the defect. Females I-2, II-4, and III-4 must all be carriers. The number of male descendants of female II-3 is insufficient to determine whether or not she is a carrier.

16b. The odds for the various cases are as follows: autosomal recessive, both parents heterozygous, 1 in 2; recessive X-linked, heterozygous mother, 1 in 2 for male children only.

16c. An autosomal dominant trait can be maintained in the population only if the defect is minor with respect to reproductive performance or if the defect, although severe, is of late onset. The latter is true of Huntington's disease.

For further study, see text section "Genetic Mapping of Mutations," p. 269

17. Gene a and gene b are on the same chromosome. The homozygous father can produce only one type of gamete, a2b2, regardless of the location of the genes. If the genes were on separate chromosomes, the mother would produce four types of gametes—a1b1, a1b2, a2b2, and a2b2—yielding offspring with four different genotypes. However, the finding that none of the offspring were either homozygous or heterozygous for both genes indicates that the mother produced only two types of gametes: a1b2 and a2b1. Since the maternal a and b genes did not assort independently; they must be "linked," that is, on the same chromosome.

For further study, see text section "Genetic Mapping of Mutations," pp. 269-270.

18. Polytene chromosomes from the salivary gland of Drosophila melanogaster, which are produced by a DNA amplification process, each contain ≈1000 copies of the DNA duplex. Mitotic chromosomes from human skin fibroblasts are unamplified. Hence the sensitivity of the in situ chromosome hybridization technique will be much higher with D. melanogaster polytene chromosomes than with human mitotic chromosomes. Even though FiberFISH uses only a single copy of a chromosome, the location of a probe can be mapped with

greater precision (to within about 400 kbp) compared to hybridization to mitotic chromosomes, where the resolution is about 3 Mb.

For further study, see text section "Molecular Cloning of Genes Defined by Mutation," pp. 274-275 and Figure 8-23. See also Figures 8-21 and 8-22.

19. In chromosome walking, a probe prepared from one end or the other of the starting segment is used to reprobe the library. For the second probe to react with a new DNA segment, there must be partial overlap of the DNA fragments included in the library. In the absence of overlap, the probe would only react with the starting segment. DNA fragments produced from incomplete restriction-enzyme digestion, digestion with two or more restriction enzymes, or sequence nonspecific techniques such as shearing by sonication will have extensive overlaps.

For further study, see text section "Molecular Cloning of Genes Defined by Mutation," pp. 275-276 and Figure 8-24.

9 Molecular Structure of Genes and Chromosomes

PART A: *Chapter Summary*

The genomes of vertebrates are complex and contain both coding and noncoding regions. This chapter describes the organization of both nuclear and organellar genomes of higher eukaryotes and the packaging of DNA into chromosomes. A gene is defined as the entire DNA sequence required for synthesis of a functional protein or RNA molecule and can exist as a simple or complex transcription unit. Only about 5% of the human genome is thought to be transcribed into RNA. Much of the genomes of higher eukaryotes contain nonfunctional DNA, some of which exists as repetitive DNA. Differences in the structural organization of this repetitive DNA are the basis for identifying individuals by a technique known as DNA fingerprinting. The size and coding capacity of organellar DNA varies among different organisms. In addition, mitochondrial genes of mammals, invertebrates, and fungi use a different genetic code for translation of mitochondrial proteins.

The genome is not static. There are mobile elements such as insertion sequences, transposons, and retrotransposons, which can move around the genome. There are two major classes of mobile elements: those that move as a DNA intermediate and those that move as an RNA intermediate. Most mobile elements encode the enzymes, e.g., transposase or reverse transcriptase, necessary for mediating the transposition event. Other changes in the organization of the genome are a result of DNA amplification and DNA recombination events.

The DNA in the nucleus is packaged into a highly ordered structure. DNA is first wrapped around a histone core to form chromatin. Chromatin is then coiled into a solenoid structure, which is attached to a chromosome scaffold composed of nonhistone proteins. The DNA-protein scaffold is further compacted to form a metaphase chromosome. Differences in the localized organization of the condensed chromatin produce unique banding patterns that are useful for chromosome identification. Artificial yeast chromosomes have been constructed for cloning millions of base pairs of DNA.

PART B: *Reviewing Concepts*

9.1 *Molecular Definition of a Gene*
(lecture date _____)

1. What are the differences between monocistronic and polycistronic RNAs?

2. In the 1940s, the conventional wisdom was "one gene, one enzyme." Discuss the validity of this phrase in terms of current knowledge about DNA structure.

9.2 *Chromosomal Organization of Genes and Noncoding DNA*
(lecture date _____)

3. Describe the reassociation kinetics typically seen for mammalian DNA.

4. What is the molecular basis for identification of individuals by DNA fingerprinting?

9.3 *Mobile DNA* (lecture date _____)

5. Describe the structural features of bacterial insertion sequences and transposons.

6. Bacterial mobile DNA elements can move in the genome by nonreplicative or replicative transposition. Describe the model for nonreplicative transposition of bacterial insertion sequences.

7. Describe the structural features of eukaryotic viral and non viral retrotransposons.

9.4 *Functional Rearrangements in Chromosomal DNA* (lecture date _____)

8. Describe the mechanism for phase variation in *Salmonella typhimurium*.

9. Describe how DNA is reorganized to form a functional κ-light chain gene.

9.5 *Organizing Cellular DNA into Chromosomes* (lecture date _____)

10. Compare the mechanisms by which bacteria and eukaryotic cells compact DNA by neutralizing the negatively charged phosphate groups on the DNA.

11. What role does acetylation/deacetylation of histones play in regulating transcriptional activity of genes?

9.6 *Morphology and Functional Elements of Eukaryotic Chromosomes* (lecture date _____)

12. Describe the different levels of packaging of DNA to form a metaphase chromosome.

13. What are the three functional elements required for replication and stable inheritance of chromosomes?

9.7 *Organelle DNAs* (lecture date _____)

14. Describe the size and organization of human mitochondrial DNA.

15. The genetic code was once considered to be a universal code. Explain why this is no longer true.

PART C: *Analyzing Experiments*

9.1 *Molecular Definition of a Gene*

16. (●●) The α-tropomyosin gene contains multiple exons and polyadenylation signals (indicated by an A). As a result of alternative splicing and polyadenylation, the primary transcript can be processed into multiple mRNAs, of which three are shown in Figure 9-1. These alternatively processed mRNAs are expressed specifically in striated muscle, smooth muscle, or brain.

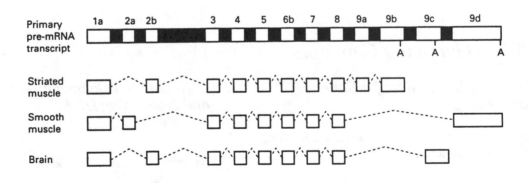

Figure 9-1

a. Is the α-tropomyosin gene classified as a simple or complex transcription unit? What is the advantage of having a gene that can be processed in multiple ways?

b. A mutation in the α-tropomyosin gene results in the loss of functional α-tropomyosin in striated muscle, smooth muscle, and brain. Where is a likely site for this mutation?

c. Another mutation in the α-tropomyosin gene results in the loss of functional α-tropomyosin mRNA in smooth muscle only. Where is a likely site for this mutation?

9.2 Chromosomal Organization of Genes and Noncoding DNA

17. (•••) Heating causes dissociation of double-stranded DNA and is accompanied by an increase in ultraviolet absorbance by the DNA bases. This phenomenon is known as DNA hyperchromicity. Figure 9-2 shows melting curves for three different calf thymus DNA samples: a freshly prepared native DNA, reassociated intermediate-repeat DNA, and reassociated single-copy DNA. The total amount of DNA in each sample is the same.

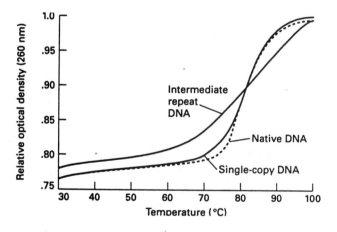

Figure 9-2

a. Why does the intermediate-repeat DNA have a higher absorbance value at 37°C than either the single-copy or native DNA?

b. As the temperature is increased, the DNA dissociates, as evidenced by hyperchromicity, in all the samples. However, denaturation occurs both at lower temperatures and over a much broader temperature range for the intermediate-repeat sample than for the other two samples. Why?

c. Based on the DNA melting profiles in Figure 9-2 does the intermediate-repeat fraction consist of exact DNA copies?

9.3 Mobile DNA

18. (••) In maize, the C-locus makes a factor required for synthesis of a purple aleurone pigment that colors the kernels (a) as shown in Figure 9-3. In this figure the dark regions represent the purple-colored regions. Insertion of the transposable elements, Ds, into the C-locus inactivates the C-locus thus making the kernels colorless (b). The Ds element is unable to transpose by itself, requiring the presence of the Ac element for transposition. After introduction of the Ac element, kernels containing both colorless and purple spots are seen (c and d).

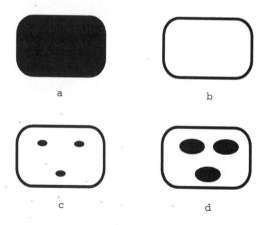

Figure 9-3

a. How do kernels containing both colorless and purple spots arise?

b. How can you explain the difference in size between the purple spots seen on kernels c and d?

c. Suppose that you have a different maize plant that contains kernels with 20 purple spots. What can you conclude about the frequency of transposition for the maize plant with kernels containing 3 spots versus the maize plant with kernels containing 20 spots?

9.5 Organizing Cellular DNA into Chromsomes

19. (••) In an electron micrograph of a human chromosome spread, you observe a thick fiber with a length of about 900 nm and an apparent diameter of 30 nm, which is expected for the solenoid structure of condensed chromatin.

   a. What is the length in base pairs of the double-helical DNA present in this fiber? Assume, for simplicity, that there is one helical turn of the solenoid per 30 nm along the fiber.

   b. What would be the effect on fiber length and diameter of preparing the chromosome spread under low-salt and low-magnesium conditions? Calculate the length of a chromatin fiber prepared under these conditions corresponding to the thick fiber described above.

9.6 Morphology and Functional Elements of Eukaryotic Chromosomes

20. (•••) Histone genes in *Drosophila* are arranged as a tandem repeat of clustered H1, H2A, H2B, H3, and H4 genes. About 100 copies of this repeat are found at the cytological 39DE locus of the polytene chromosome. Laemmli and colleagues investigated how this repeat is bound to the nuclear scaffold by determining whether histone gene fragments were retained in a scaffold preparation following digestion of the preparation with a mixture of restriction enzymes. In brief, a scaffold preparation was obtained from nuclei containing polytene chromosomes, digested with restriction enzymes, and pelleted; the DNA from the pellet (P) and supernatant (S) was then analyzed by Southern blotting using a radio-labeled histone-gene repeat probe. Typical results are shown in Figure 9-4. In these experiments, the nuclei were incubated with Cu^{2+} or at 37°C to stabilize the scaffolds. The actual nuclear scaffolds were then prepared by extracting the pellet fraction with lithium diiodosalicylate (LIS).

   a. In this experimental protocol, what result indicates that a histone-repeat fragment is bound to the scaffold?

   b. What can you conclude from the data in Figure 9-4 about the location of the scaffold-binding site in the histone-gene repeat?

   c. How could you locate more precisely the scaffold-binding site in the histone-gene repeat?

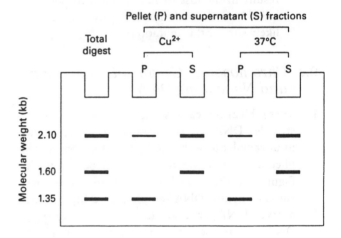

Figure 9-4

Answers

1. Polycistronic mRNAs, which are common in prokaryotes, consist of mRNAs that encode several polypeptides. Ribosomes initiate translation at each of the protein coding regions in the mRNA. Each protein coding region is called a cistron and hence the name polycistronic mRNA. In contrast, most eukaryotic transcription units are monocistronic and produce mRNAs that code for only one protein. Monocistronic mRNAs have only one ribosome-binding site per mRNA.

   For further study, see text, p. 295.

2. The concept of *one gene, one enzyme*, which was based largely on the results of classical genetic analyses, has not stood the test of time, particularly for eukaryotes. The development of more sophisticated techniques for studying the structure of genes and the structure and function

of proteins led to the discovery that although some genes code for enzymes, others code for structural proteins without enzymatic activity or for RNA molecules, which are not translated into proteins. These phenomena are true of prokaryotes as well as eukaryotes. In eukaryotes, complex transcription units encode multiple polypeptides, which are generated by alternative processing of a single primary transcript. The use of alternative poly A sites or splice sites can lead to the synthesis of multiple proteins from a single gene.

For further study, see text, pp. 295–296.

3. In DNA reassociation experiments, approximately 10–15% of mammalian DNA reassociates at a rapid rate. This rapidly reassociating DNA consists of short sequences (5- to 10-base pairs) repeated in long tandem arrays. Another 25–40% of mammalian DNA reassociates at an intermediate rate. This DNA fraction consists of moderately repeated DNA that is located at interspersed regions of the genome. The final 50–60% of mammalian DNA reassociates at a slow rate. This fraction consists primarily of single copy DNA, i.e., DNA sequences that are present only once in the genome.

For further study, see text, p. 301.

4. The identification of individuals based on DNA fingerprinting depends on differences in the length of tandem DNA arrays. DNA sequences of 15- to 100-base pairs are present as tandem arrays in the genome and are known as microsatellites. The number of repeat units, and thus the length of simple sequence tandem arrays, varies between individuals. The difference in the length of these tandem arrays can be detected by Southern blot analysis. In this method a DNA sample is digested with a restriction enzyme that usually does *not* cut within the tandemly repeated satellite array. A DNA probe known to hybridize with these repeat sequences is then used to reveal the resulting differences in lengths of the tandem arrays.

For further study, see text, p. 302 and Figure 9-9.

5. Bacterial insertion sequences or IS elements range from 1- to 2-kilobases in length and encode a transposase necessary for transposition. The transposase is flanked by approximately 50 base pair inverted repeats. Bacterial transposons are larger than IS elements and contain one or more protein coding genes (e.g., antibiotic resistance genes) in addition to the genes required for transposition. Two copies of the same IS element flank the antibiotic resistance gene.

For further study, see text, pp. 304–307 and Figures 9-11 and 9-13.

6. Non-replicative transposition of bacterial IS elements is mediated by a transposase. The transposase, which is encoded in the IS element, cleaves the donor DNA molecule between the terminal direct repeats and the inverted repeats and thus excises the IS element. The same transposase also makes staggered cuts in the target DNA. The 3' termini of the IS element are ligated to the staggered ends of the target DNA by the action of the transposase. The single-stranded gaps are filled in by the cellular DNA polymerase, and the final phosphodiester bond is synthesized by the cellular DNA ligase. The resulting double stranded molecule contains an insertion element flanked by direct repeats.

For further study, see text, p. 306 and Figure 9-12.

7. Viral retrotransposons contain a central protein coding region flanked by two 250-600 base pair long terminal repeats (LTR). The leftward LTR functions as a promoter that directs host-cell RNA polymerase to initiate transcription. The rightward LTR provides the polyadenylation site. The two enzymatic activities encoded in the retrotransposon are reverse transcriptase for synthesizing DNA from an RNA template and integrase for inserting the double stranded DNA into the genome. Non-viral retrotransposons lack LTRs and have an A/T-rich region at one end. Examples of these are long interspersed elements (LINES) and short interspersed elements (SINES).

For further study, see text, pp. 307–308 and Figures 9-14 and 9-18.

8. The mechanism for phase variation in *Salmonella typhimurium* involves a DNA inversion event. In phase I, the H2 flagellar protein is synthesized along with a repressor protein that represses transcription of the H1 flagellar protein gene. A DNA segment that is located adjacent to the flagellar H2 operon functions as a transcriptional promoter. In phase II, the H2 promoter sequence is inverted by a site-specific recombinase. In this orientation, the H2 flagellar protein and the repressor of H1 are not synthesized, which leads to the synthesis of only H1 flagellar protein. For further study, see text, p. 314 and Figure 9-20.

9. The κ-light chain gene in germ line DNA consists of approximately 100 leader (Lκ) and variable (Vκ) units followed by five joining (Jκ) segments and one constant (C) segment. During the formation of a functional kappa gene, one LκVκ unit joins to one Jκ segment. This recombination event results in the deletion or inversion of the intervening DNA sequence depending upon the relative orientation of the Lκ-Vκ unit with the Jκ segment. Once the Vκ-Jκ joining has occurred, the gene can be transcribed.

   For further study, see text, p. 316 and Figures 9-23 and 9-24.

10. In the bacterium *E. coli*, the negatively charged phosphate groups of DNA are neutralized by association with the positively charged polyamines, spermine and spermidine. In addition, numerous small proteins, such as H-NS bind to and compact DNA. In eukaryotes, DNA is associated with positively charged proteins called histones. This DNA-histone complex is known as chromatin.

    For further study, see text, pp. 320–321.

11. Histones contain several positively charged lysine groups that interact and neutralize the negatively charged phosphate groups on DNA. Acetylation of the lysine ε-amino group neutralizes the positive charge on the lysine and disrupts the interactions between histones and DNA phosphate groups. In this manner, histone acetylation promotes the formation of an open chromatin structure, which is a feature of actively transcribed genes. In an opposite fashion, unacetylated histones promote chromatin condensation and the formation of nontranscribed DNA. For further study, see text, pp. 323–324.

12. DNA is first wrapped around histone cores to form chromatin. The chromatin is next packed into a solenoid arrangement to form the 30-nm chromatin fiber. Long loops of 30-nm fibers are anchored to the nonhistone protein chromosome scaffold at scaffold-associated regions or matrix attachment regions. The scaffold is then folded into a helix to form a highly compacted structure characteristic of a metaphase chromosome.

    For further study, see text, pp. 325–326 and Figure 9-35.

13. The three functional elements required for replication and stable inheritance of chromosomes are an origin of DNA replication, the centromere and two telomeres. In yeast, the origin of DNA replication is known as an autonomously replicating sequence. The centromere is required for proper segregation of chromosomes during mitosis and is the attachment site for the mitotic spindle. The telomeres are specialaized structures at the ends of linear chromosomes that are necessary to prevent shortening of the ends of chromosomes during each round of DNA replication.

    For further study, see text, pp. 329–331.

14. The human mitochondrial genome is a circular DNA molecule containing 16,569 base pairs. It encodes two mitochondrial ribosomal RNAs, 22 tRNAs used to translate mitochondrial mRNAs, and 13 proteins. Mitochondrial DNA lacks introns and contains no long noncoding sequences.

    For further study, see text, p. 334 and Figure 9-44.

15. The genetic code used in animal and yeast mitochondria is different from the standard genetic code used for all prokaryotic and eukaryotic nuclear genes. For example, UGA codes for a stop codon in the standard genetic code, whereas in

human and yeast mitochondria, UGA codes for a tryptophan. Thus within the same human cell, UGA is read as a stop codon in the nucleus but as a tryptophan codon in the mitochondria.

For further study, see text, p. 335 and Table 9-4.

16a. The α-tropomyosin gene would be considered a complex transcription unit because multiple mRNAs are generated as a result of alternative splicing and polyadenylation usage. One advantage of a complex transcription unit is the generation of diversity. From a single gene, multiple mRNAs and thus multiple proteins can be synthesized.

16b. A mutation in the promoter/regulatory region of this gene, i.e., upstream of exon 1a would result in loss of α-tropomyosin mRNA transcription in all cells. A second possibility is a mutation that changes the amino acid sequence in a shared exon, i.e., exons 1a, 3, 4, 5, 6b, 7 or 8. A third possibility is a mutation that alters a splice site for a shared exon, resulting in an incorrectly spliced mRNA. These mutations could result in the synthesis of a nonfunctional α-tropomyosin in all cells.

16c. A mutation in an exon that is present only in smooth muscle α-tropomyosin mRNA could lead to loss of functional α-tropomyosin specifically in smooth muscle. In this case, exons 2a and 9b are unique to smooth muscle α-tropomyosin mRNA. A mutation that changes the amino acid sequence or a splice site in exons 2a or 9b could result in loss of functional α-tropomyosin only in smooth muscle cells.

17a. The absorbance of a DNA sample is inversely proportional to the extent of base pairing: the less extensive the base pairing, the higher the absorbance. The higher absorbance value at 37°C for the reassociated intermediate-repeat DNA thus indicates that complete base pairing is not occurring along all portions of the DNA. During the renaturation process, rapid intermolecular and intramolecular base pairing of the most repeated DNA may result in a DNA network that prevents complete base pairing.

17b. The denaturation of the reassociated intermediate-repeat DNA at the lower temperature is another indication of its less extensive, less complete base pairing. The broader temperature range for the denaturation of this sample is an indication of variable degrees of base pairing in different portions of the DNA. If all of the DNA were fully base-paired along its length, then the denaturation would occur over a very narrow temperature range as occurs with the renatured single-copy DNA and native DNA samples.

17c. The broad temperature range over which the reassociated intermediate-repeat DNA denatures strongly suggests that the DNA does not consist of exact copies. If the copies were exact, the DNA would be fully base-paired and denature over a narrow temperature range.

18a. Insertion of the Ds transposable element into the C locus leads to loss of a factor required for synthesis of a purple aleurone pigment. In the presence of the Ac element, the Ds element is excised from the C locus, which allows re-synthesis of the purple aleurone pigment. Therefore, a maize kernel that contains both colorless and purple spots is due to a combination of cells that still contain the Ds transposable element (colorless) and cells that lack Ds (purple) at the C locus.

18b. The size of the purple spots is a reflection of the time in development during which the Ds element was excised. In the case of kernel c, small purple spots indicate that excision of Ds occurred late in development. The clone of cells without the Ds element is small and thus the size of the pigmented area is small. In contrast, the excision event occurred earlier in development in kernel d, which generates larger purple spots.

18c. The number of purple spots is a reflection of the frequency of the excision (transposition) event. The maize plant that contains kernels with 20 spots shows a much higher frequency of excision of Ds from the C locus than the maize plant with kernels containing three spots.

19a. There are six nucleosomes per helical turn of the solenoid structure, and one helical turn of the solenoid corresponds to slightly less than 30 nm along the length of a chromatin thick fiber. Assuming, for simplicity of calculation, one helical turn per 30 nm, then there are 6 nucleosomes per 30-nm stretch of thick fiber. A 900-nm-long, thick fiber thus has 30 solenoid turns (900 nm divided by 30 nm/turn) and contains 180 nucleosomes (6 nucleosomes/turn × 30 turns). The DNA content of each human nucleosome plus the linker DNA connecting it to adjacent nucleosomes is about 200 bp. This thick fiber thus contains 36,000 bp of DNA: (200 bp/nucleosome) × (180 nucleosomes/900-nm thick fiber).

19b. If the chromosome spread was prepared under low-salt and low-magnesium conditions, the chromosome fiber seen would be a thin, 10-nm-diameter fiber with a "beads-on-a-string" appearance, as illustrated in Figure 9-5. Each individual nucleosome has a length of 10 nm with 140 bp of DNA wound about it and 60 bp of DNA linking it to adjacent nucleosome beads. As discussed in Chapter 4 of MCB, B-form DNA, the standard double-helical form of DNA, makes a complete helical turn every 3.4 nm, and each turn contains 10 bp. Thus 60 bp of linker DNA corresponds to a linear stretch of 20.4 nm (60 bp × 3.4 nm/10 bp). For each "bead-on-a-string" repeat (nucleosome plus linker DNA), the fiber length should be 30 nm: 10 nm/nucleosome + 20 nm/linker DNA. Thus the beads-on-a-string form of a 900-nm-long thick fiber, which contains 180 nucleosomes, should be 5400-nm long: 30 nm/repeat × 180 repeats. In other words, an extended chromatin fiber is 6 times longer than a thick, condensed chromatin fiber (5400 nm divided by 900 nm).

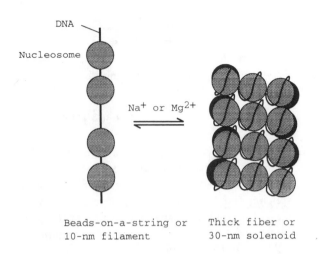

Figure 9-5

20a. In this experiment, the nuclear scaffold and any associated DNA is located in the pellet fraction; however, free DNA and DNA fragments are located in the supernatant fraction. Thus detection of a histone-repeat restriction fragment in the pellet indicates that the fragment is bound to the scaffold.

20b. The Southern-blot pattern suggests that the 1.35-kb fragment is specifically bound to the scaffold preparation (the pellet fraction), as this fragment is found exclusively in the pellet fraction. The 1.60-kb fragment is completely released into the supernatant fraction by the restriction endonuclease digestion, as is almost all of the 2.10-kb fragment. Hence the 1.35-kb fragment, the only fragment preferentially retained with the scaffold preparation, appears to contain the scaffold-binding site in the histone gene repeat.

20c. The use of additional or alternative restriction endonucleases to produce smaller fragments would increase the precision of this determination.

10

Regulation of Transcription Initiation

PART A: *Chapter Summary*

The regulation of gene expression can be controlled at a number of different levels. This chapter focuses on an analysis of the regulation of transcription initiation in both prokaryotic and eukaryotic systems. The Jacob-Monod model for inducible regulation of the *E. coli lac* operon is one of the best-understood examples of gene control, involving the interaction of a repressor, inducer, operator, and promoter. Eukaryotic gene control involves many of the same elements as bacterial gene control, only more complex.

Eukaryotic genes are regulated by the interaction of transcription factors with cis-acting regulatory elements such as promoter-proximal elements and enhancers. The identification of these regulatory elements and their DNA binding proteins was facilitated using a number of techniques such as the electrophoretic mobility shift assay, DNase I footprinting, 5' deletion analysis, linker scanning mutation analysis, and sequence-specific DNA affinity chromatography. Transcription factor activators and repressors consist of two modular domains with distinct functions: a DNA binding domain and an activation/repression domain. The DNA binding domain can be classified into a number of different structural motifs such as the homeodomain, zinc-finger structure, and leucine zipper.

Eukaryotic transcriptional control operates at three levels: modulation of the levels or activities of activators and repressors, changes in chromatin structure directed by activators and repressors, and influence of activators and repressors on the transcription initiation complex. The transcription initiation complex for pre-mRNA genes is a multiprotein complex that includes RNA polymerase II and a number of general transcription factors. This multiprotein complex consists of 60–70 polypeptides with a total mass of approximately 3 MDa, almost the size of a eukaryotic ribosome. Transcription initiation at RNA polymerase I and polymerase III promoters is similar to but not identical with that at RNA polymerase II promoters.

PART B: *Reviewing Concepts*

10.1 *Bacterial Gene Control: The Jacob-Monod Model* (lecture date _____)

1. Describe the molecular events that occur at the *lac* operon when *E. coli* cells are shifted from a glucose-containing medium to a lactose containing medium.

2. Describe the experimental evidence that O^c mutations are cis-acting.

10.2 *Bacterial Transcription Initiation* (lecture date _____)

3. Regulation of the lac operon is under both negative and positive control. Describe how these two transcriptional control mechanisms regulate expression of the *lac* operon in the presence of glucose and/or lactose.

4. Control of the *E. coli PhoR/PhoB* genes is controlled by a two-component regulatory system. Describe how this two-component regulatory system functions.

10.3 *Eukaryotic Gene Control: Purposes and General Principles*
(lecture date _____)

5. Compare the structure and function of the three eukaryotic RNA polymerases.

10.4 *Regulatory Sequences in Eukaryotic Protein-Coding Genes*
(lecture date _____)

6. Describe the three types of eukaryotic promoters.

7. How can linker scanning mutation analysis be used to map the location of transcription-control elements?

10.5 *Eukaryotic Transcription Activators and Repressors* (lecture date _____)

8. Describe the experimental evidence that shows that transcriptional activators have separable DNA binding and activation domains.

9. What is the functional advantage of heterodimeric transcription factors?

10.6 *RNA Polymerase II Transcription-Initiation Complex*
(lecture date _____)

10. Describe the sequence of events that occur during formation of a transcription initiation complex on a TATA box promoter.

10.7 *Molecular Mechanism of Eukaryotic Transcriptional Control*
(lecture date _____)

11. Describe the general mechanism of yeast mating-type switching. In what way is this mechanism similar to a cassette tape player?

12. Describe the mechanism for gene silencing at yeast telomeres.

10.8 *Other Transcription Systems*
(lecture date _____)

13. Describe transcription initiation for RNA polymerase I and III genes.

14. Describe how the archaebacteria are similar to both eukaryotes and prokaryotes in their mechanism of gene transcription.

PART C: *Analyzing Experiments*

10.4 *Regulatory Sequences in Eukaryotic Protein-Coding Genes*

15. (••) One of the first promoter regions to be analyzed by linker scanning mutations was that of the thymidine kinase (*tk*) gene of herpes simplex virus. The results of such a study for the −100 region of the gene are shown in Figure 10-1. Each of the boxes in the figure shows the position of a clustered region of 6 to 10 nucleotide substitutions.

   a. Based on these results, draw a map showing the position of specific nucleotide regions important for the efficient transcription of the *tk* gene.

   b. What type of control element is represented by each of these transcriptionally important regions?

   c. Which of these transcriptionally important regions is likely to contain two neighboring control elements?

10.1 *Bacterial Gene Control: The Jacob-Monod Model*

16. (••) One of the classic studies contributing to our understanding of the *lac* operon was the PaJaMo experiment. In this experiment, diploid *E. coli* cells (merozygotes) were formed by conjugation of $I^+ Z^+$ (donor) cells with $I^- Z^-$ (recipient) cells in the absence of inducer. The levels of β-galactosidase activity in the merozygotes were monitored as a function of time and of inducer addition. The basic experimental protocol and the results observed are summarized in Figure 10-2.

   a. Explain why an increased rate of β-galactosidase synthesis was observed initially in the diploid bacteria and why at later times inducer was required for rapid β-galactosidase synthesis.

   b. What can you conclude about the nature of the inducer from this experiment?

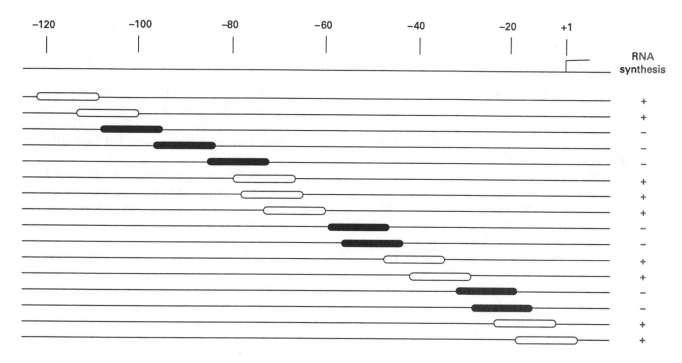

Figure 10-1

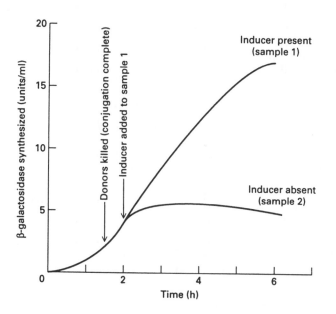

Figure 10-2

10.2 *Bacterial Transcription Initiation*

17. (●●) The NFκB transcription factor binds to an enhancer of the kappa light-chain gene, an enhancer on the human immunodeficiency virus (HIV) genome, and an upstream sequence of the gene encoding the a chain of the interleukin-2 receptor. Binding of NFκB confers transcriptional activity and phorbol ester inducibility to genes controlled by these cis-acting elements. This factor can exist in active and inactive forms. Baeuerle and Baltimore have used electrophoretic mobility shift analysis (EMSA) to investigate the nature of these two forms. EMSA can detect the ability of a transcription factor to retard the mobility of enhancer DNA in a gel.

In one such experiment, isolated NFκB factor was incubated with [32P]-labeled enhancer DNA in the presence or absence of varying amounts of complete cytosol or NFκB-depleted cytosol from pre-B cells. The migration of the labeled DNA in the gel following incubation is shown in Figure 10-3. Free DNA migrates rapidly and is found at the bottom of the gel. The addition of increasing amounts of cytosol or depleted cytosol to the incubation mix are indicated by the numbers 1 → 4 at the top of the figure. As controls, incubation is with cytosol or depleted cytosol alone (lanes 1 and 2).

a. Based on lanes 1 and 2 of the EMSA pattern, what is the effect of NFκB on enhancer DNA migration?

b. Does cytosol have any apparent effect on the migration of enhancer DNA incubated with purified NFκB?

c. Does NFκB-depleted cytosol have any apparent effect on the migration of enhancer DNA incubated with purified NFκB? If so, does depleted cytosol appear to contain an activator or inhibitor of NFκB activity?

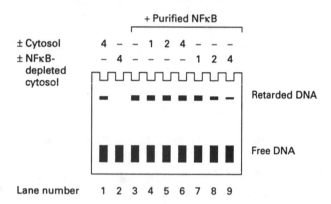

Figure 10-3

10.5 Eukaryotic Transcription Activators and Repressors, 10.7 Molecular Mechanisms of Eukaryotic Transcriptional Control

18. (•) You have constructed a set of plasmids containing a series of nucleotide insertions spaced along the length of the glucocorticoid-receptor gene. Each insertion encodes three or four amino acids. The map positions of the various insertions in the coding sequence of the receptor gene are as follows:

The plasmids containing the receptor gene can be functionally expressed in CV-1 and COS cells, which contain a steroid-responsive gene. Using these cells, you determine the effect of each of these insertions in the receptor on the induction of the steroid-responsive gene and on binding of the synthetic steroid dexamethasone. The results of these analyses are summarized in Table 10-1.

a. From this analysis, how many different functional domains does the glucocorticoid receptor have? Indicate the position of these domains on the insertion map.

b. Which domain is the steroid-binding domain?

c. How could you determine which of the domains is the DNA-binding domain?

Insertion	Induction	Dexamethasone binding
A	++++	++++
B	++++	++++
C	++++	++++
D	0	++++
E	0	++++
F	0	++++
G	++++	++++
H	++++	++++
I	+	++++
J	++++	++++
K	0	++++
L	0	++++
M	0	++++
N	+	++++
O	++++	++++
P	++++	++++
Q	0	0
R	0	0
S	0	0

Table 10-1

10.3 *Eukaryotic Gene Control: Purposes and General Principles*

19. (•) You have cloned a 10-kb piece of genomic DNA that includes the full transcription unit for your prize protein. You wish to determine the approximate locations of the initiation and termination sites for the transcription unit by nascent-chain analysis of labeled RNA. The restriction map of this fragment is shown in Figure 10-4.

    a. You test each of these restriction fragments for hybridization to a population of nascently labeled RNA chains. Your results are shown in Table 10-2. Where are the initiation and termination sites for the transcription unit located?

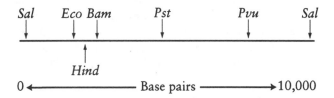

Figure 10-4

    b. What is the maximum length of the mRNA that could be coded by this piece of genomic DNA?

    c. How could the precision of the assignment of the initiation and termination sites be improved?

Restriction fragment	Hybridization
Sal-Sal	+
Sal (left end)-*Eco*	-
Eco-Bam	-
Bam-Pst	+
Pst-Pvu	+
Pvu-Sal (right end)	-

Table 10-2

10.4 *Regulatory Sequences in Eukaryotic Protein-Coding Genes, 10.8 Other Transcription Systems*

20. (•••) A template competition assay for transcription factor specificity includes the following steps: (1) gene A is incubated with limiting amounts of transcription factor, (2) competing gene B is added, and (3) transcription of each gene is assayed by gel electrophoresis of the RNA transcripts. Isolated human transcription factors for RNA polymerase III genes were assayed with this procedure, using a 5S-rRNA gene with an insert (termed long gene) and a wild-type 5S gene. The gel patterns of the reaction products (long rRNA, wild-type rRNA) are shown diagrammatically in Figure 10-5; the components incubated in step 1 are indicated at the top (A, B, and C refer to TFIII A, B, and C, and III to RNA polymerase III). In all reaction mixtures except that shown in gel lane 1, the genes were added sequentially with the long gene added first. In the reaction shown in lane 1, the two genes were added simultaneously.

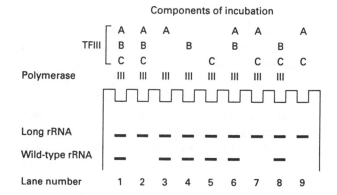

Figure 10-5

    a. What conclusions regarding the role of TFIIIA, B, and C and RNA polymerase III in the formation of protein-DNA complexes are suggested by these data?

    b. If TFIIIA acted catalytically rather than stoichiometrically, would further TFIIIA be required for transcription of the long gene added to a reaction mix containing excess TFIIIB and C, RNA polymerase, wild-type 5S gene, and limiting amounts of TFIIIA?

c. When Roeder and colleagues performed the experiment described in part (b), they found that more TFIIIA was required for transcription of the long gene. In parallel experiments with TFIIIC, similar results were obtained. Based on these data, what is the role of TFIIIA and C in the transcription of 5S-rRNA genes?

Answers

1. In the presence of glucose the *lac* operon is turned off. The *lac* repressor is synthesized and bound to the operator, which prevents binding of RNA polymerase to the promoter and subsequent activation of the *lac* operon. When *E. coli* cells are shifted into a lactose-containing medium in the absence of glucose, the lactose binds to the *lac* repressor and prevents its binding to the operator. In the absence of *lac* repressor binding to the operator, RNA polymerase can bind to the promoter and turn on the *lac* operon. For further study, see text, p. 342 and Figure 10-2.

2. *E. coli* cells can contain two copies of the *lac* operon, one of which is a chromosomal copy and the other which is on a plasmid. In an *E. coli* cell that contains the genotype I⁺OᶜZ⁻ on its chromosomal copy and I⁺O⁺Z⁺ on a plasmid would be unable to synthesize β-galactosidase. In this case the repressor binds to the functional operator on the plasmid copy and inhibits the synthesis of β-galactosidase. Even though the repressor cannot bind to the operator on the chromosomal copy, no functional β-galactosidase is synthesized because the Z gene is mutant. This result demonstrates that the Oᶜ mutation only affects the *lacZ* gene on the same DNA molecule. For further study see text, p. 344 and Figure 10-3.

3. In the presence of glucose and the absence of lactose, cAMP levels are low. The *lac* repressor is bound to the *lac* operator which blocks the synthesis of *lac* mRNA. In the presence of glucose and lactose, the *lac* repressor is bound to lactose and is unable to bind to the *lac* operator. However, because cAMP is low in the presence of glucose, very little *lac* mRNA is synthesized. In the absence of glucose and the presence of lactose,

the *lac* repressor is again unable to bind to the *lac* operator because it is bound to lactose. However, cAMP levels are high in the absence of glucose. A cAMP-CAP complex forms which binds to the CAP site and stimulates transcription of *lac* mRNA. For further study, see text, pp. 352–353 and Figure 10-16.

4. In two-component regulatory systems, one protein acts as a sensor and the other protein is a response regulator. In *E. coli* the proteins PhoR (the sensor) and PhoB (the response regulator) regulate transcription of target genes in response to the free phosphate concentration. PhoR is a transmembrane protein, whose periplasmic domain binds phosphate and whose cytoplasmic domain has protein kinase activity. Under low phosphate conditions, PhoR undergoes a conformational change that activates the protein kinase activity and results in the transfer of a phosphate group to Pho B. The phosphorylated form of PhoB then induces transcription from several genes that help the cell respond to low phosphate conditions. For further study, see text, pp. 356–357 and Figure 10-21.

5. In eukaryotic cells, three different RNA polymerases catalyze the formation of different RNAs. RNA polymerase I is located in the nucleolus and synthesizes preribosomal RNA (the precursor to 28S, 5.8S, and 18S rRNAs). RNA polymerase II synthesizes messenger RNAs. RNA polymerase III synthesizes tRNAs, 5S rRNAs, and a number of small RNAs. These small RNAs include an RNA involved in RNA splicing and the 7S RNA of the signal recognition particle. All three RNA polymerases contain two large subunits and 12–15 smaller subunits. Some of these subunits are shared among the different polymerases and others are unique. For further study, see text, pp. 361–362 and Figure 10-26.

6. Three types of promoter sequences have been identified in eukaryote genes. One class of eukaryotic genes contains a highly conserved sequence located approximately 25–35 base pairs upstream of the transcription start site known as the TATA box. The TATA box acts similar to a prokaryotic promoter to position RNA polymerase II for transcription initiation. A second class of eukaryotic genes lack a TATA box and

instead contain an alternative promoter element called an initiator. The initiator sequence is not highly conserved. A final class of eukaryotic genes contains neither a TATA box nor an initiator element. These genes contain a CG-rich stretch of 20–50 nucleotides (CpG islands) approximately 100 base pairs upstream of the start site. For further study see text, pp. 365-366 and Figure 10-30.

7. In linker scanning mutation analysis, a region of DNA that contains putative regulatory elements are cloned into a plasmid upstream of a reporter gene. A series of overlapping linker scanning mutations are introduced into the DNA containing the regulatory elements. The mutations are introduced by scrambling the nucleotide sequence in a short stretch of DNA. The effects of these mutations are evaluated by assaying reporter gene expression. For further study, see text, pp. 366–367 and Figure 10-31.

8. Transcription activators were first shown to contain separable DNA binding and activation domains in yeast. A series of Gal4 mutants were constructed to determine their ability to bind to UAS$_{GAL}$, a regulatory element that contains several Gal4 binding sites, linked to a *lacZ* reporter gene. Deletion of amino acids from the N-terminal end destroyed the ability of Gal4 to bind to UAS$_{GAL}$ and to stimulate expression of *lacZ*. Gal4 mutants lacking amino acids at the C-terminal end were still able to bind UAS$_{GAL}$ but lacked the ability to stimulate expression of *lacZ*. These results localize the DNA binding domain to the N-terminal end and the activation domain to the C-terminal end. For further study, see text, pp. 372–373 and Figure 10-38.

9. Heterodimeric transcription factors increase gene control options due to an increase in the number of combinatorial possibilities. For example, assume that there are three transcription factors that exist as dimers and each one has a different DNA binding specificity. Three different combinations of heterodimers can be formed in addition to the three homodimers making a total of six different dimers. Each of these six dimers could bind to a different DNA sequence and alter transcription. For further study, see text, pp. 376–377 and Figure 10-45.

10. A number of sequential protein binding steps occurs during the formation of a transcription initiation complex. The TATA-box binding protein (TBP) included within TFIID first binds to a TATA box in the promoter. Once TBP has bound to the TATA box, TFIIB can bind. The next step is the binding of a pre-formed TFIIF and PolII complex. Then TFIIE and TFIIH bind, completing assembly of the transcription initiation complex. For further study, see text, pp. 381–382 and Figure 10-50.

11. The yeast mating-type phenotype (a or α) is determined by the presence of a protein termed mating-type factor a or α. Only one of these factors is expressed in a haploid cell, even though the genes for both are present in the genome, one at a locus called *HML* and the other at a locus called *HMR*. When at these loci, the genes are transcriptionally silent. In order to be expressed, the a or α coding sequence must be present at the *MAT* locus, which is between *HML* and *HMR*. Thus, just as a tape inserted into a cassette player can be heard, an a or α sequence inserted into the *MAT* locus can be transcribed. A haploid yeast cell switches mating type (e.g., α → a) by gene conversion, a process in which the α sequence present at *MAT* is excised and the a sequence at the silent *HML* or *HMR* locus is copied into *MAT*. The a → α switch occurs by a similar process in which the a sequence at *MAT* is lost and the silent α sequence is inserted into *MAT*. Since the mating-type sequences *HML* and *HMR* are transferred to *MAT* by DNA synthesis, they are not lost from these loci and can be passed onto the next generation of cells. For further study, see text, p. 385 and Figure 10-55.

12. The model for silencing at yeast telomeres involves the formation of a large nucleoprotein complex. First multiple copies of the protein Rap1 bind to repeated, nucleosome-free sequences at yeast telomeres. Then a multiprotein complex is formed through protein-protein interactions between the silent information regulator proteins Sir2, Sir3, and Sir4 and the hypoacetylated histones H3 and H4. As a result, the DNA in this stable nucleoprotein complex is largely inaccessible to external proteins and is silenced. For further study, see text, pp. 385–387 and Figure 10-57.

13. Transcription initiation by RNA polymerase I is mediated by a core element, which includes the start site of the pre rRNA gene and an upstream control element, located approximately 100 base pairs upstream. Two transcription factors—upstream binding factor (UBF) and selectivity factor 1 (SL1)—bind to and help stabilize the initiation complex. Transcription initiation by RNA polymerase III is mediated by internal promoter elements, termed the A box and B box, present in all tRNA genes. Two transcription factors, TFIIIC and TFIIIB, are required for transcription initiaiton of tRNA genes. An additional factor (TFIIIA) is required for transcription initiation of 5S rRNA genes. For further study, see text, pp. 397–398 and Figure 10-69.

14. Archaebacteria have a single RNA polymerase like bacteria, but a complexity equivalent to that of eukaryotic RNA polymerase. Although archaens transcribe operons into polycistronic RNAs like bacteria, archael promoters are similar to eukaryotic promoters. Furthermore, archael transcription initiation factors are homologous to eukaryotic TBP and TFIIB. Thus although transcription in archaens shares features common with both bacterial and eukaryotic systems, it is closer to the eukaryotic system. For further study, see text, p. 399.

15a. Only mutations indicated by the black boxes shown in Figure 10-6 affect RNA synthesis. These mutations fall into three regions, all of which are important for efficient transcription of the *tk* gene. These regions are centered at about -20, -50, and -90.

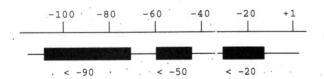

Figure 10-6

15b. Since TATA boxes generally are located within ≈25–35 bp of the start site, the control sequence centered at about -20 ought to be a TATA box. The other two sequences are likely to be promoter-proximal elements.

15c. The most upstream region is about 35–40 bp in length, whereas the other two regions are each only about 10 bp long. The longer region is the only one that is likely to contain two unresolved control elements.

16a. Since the I gene codes for repressor and the Z gene codes for β-galactosidase, the recipient I^-Z^- cells are incapable of synthesizing either repressor or enzyme; these cells do, however, have an intact operator to which repressor can bind. As the donor I^+Z^+ genome enters the I^-Z^- cells, some of the repressor bound to the donor operators is released and binds to free operators in recipient cells; transcription of the Z^+ gene on the donor genome can then proceed. Gradually, however, new repressor is synthesized in the diploid, leading to re-repression of the Z^+ gene unless inducer is added.

16b. If the explanation above is correct, then the *lac* repressor is diffusible and limiting in amount with respect to regulation of the *lac* operon. Moreover, the repressor can dissociate from the donor operator at a sufficient rate to support the initial increase in β-galactosidase synthesis observed.

17a. Removal of NFκB from the cytosol (lane 2) eliminates retardation of enhancer DNA by cytosol (lane 1). Thus NFκB retards enhancer DNA migration.

17b. The amount of NFκB in the cytosol preparation appears to be insufficient to increase the quantity of enhancer DNA retarded by the added purified NFκB (lanes 3–6).

17c. Addition of depleted cytosol to the incubation mixtures containing purified NFκB results in a decrease in the quantity of enhancer DNA that is retarded (lanes 3, 7–9). This observation suggests that depleted cytosol contains an inhibitor of NFκB activity.

18a. The data in Table 10-1 show that insertions in four different regions of the receptor coding sequence cause decreased induction. This finding suggests that the glucocorticoid receptor has four separate functional domains. These four domains correspond to the following insertions: domain 1 = insertion D, E, and F; domain 2 = insertion I; domain 3 = insertion K, L, M, and N; and domain 4 = insertion Q, R, and S. See Figure 10-7.

18b. Only insertions Q, R, and S produce receptor proteins with decreased steroid-binding ability. Therefore, the region of the protein corresponding to these insertions is the steroid-binding domain.

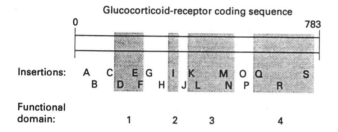

Figure 10-7

18c. The data indicate that insertions in three receptor domains block induction without affecting steroid binding. An EMSA-type assay could be used to determine whether changes in any of these three functional domains affect binding of the dexamethasone-receptor complex to DNA. Alternatively, the DNA-binding domain could be inferred by comparing the sequence of glucocorticoid receptor with that of known DNA-binding domains in other steroid hormone receptors.

19a. As expected, fragment A is positive for nascent-chain hybridization, whereas fragments B and C are negative; therefore the initiation site must be to the right of the *Bam* site. Since the *Pvu-Sal* fragment (F) also is negative, the transcript does not include the *Pvu-Sal* region. These results indicate that the initiation site is somewhere between the *Bam* site and the *Pvu* site. Because the orientation of the transcript with respect to

the restriction map is not known, the initiation site could be either towards *Bam* or *Pvu*. By the same reasoning, the termination site must be somewhere between the *Bam* site and the *Pvu* site. Whether it is closer to *Bam* or *Pvu* cannot be determined from the data.

19b. The maximum length of mRNA is the distance between the *Bam* and *Pvu* sites.

19c. Mapping of additional restriction fragments and hybridization experiments with them would improve the determination of mRNA length. Establishment of mRNA orientation relative to the restriction map would allow assignment of the initiation and termination sites relative to left and right on the map.

20a. In this assay, formation of a stable complex between the long gene and any of the transcription factors during the first incubation would prevent expression of the wild-type gene; thus only long rRNA would be observed in the gel. Lanes 3–5, which reveal both products, indicate that no factor alone forms a stable complex. Lanes 2 and 7 suggest that a stable, bound complex of TFIIIA and C is formed when these transcription factors are incubated with 5S-rRNA genes. Formation of this complex does not require RNA polymerase III (lane 9).

20b. If TFIIIA acted catalytically in the formation of a stable complex of TFIIIC with 5S-rRNA genes, then TFIIIA should be available in the incubation mix and would be able to catalyze the formation of a TFIIIC complex with added DNA. No additional A should be needed.

20c. These data indicate that neither TFIIIA nor TFIIIC acts catalytically. As complex formation is stable and DNA dependent, the most likely role of the transcription factors is to form a stable, DNA-bound intermediate in the formation of a RNA polymerase III transcription-initiation complex.

11

RNA Processing, Nuclear Transport, and Post-Transcriptional Control

PART A: *Chapter Summary*

The regulation of most genes occurs at the first step in gene expression, namely, initiation of transcription. However, once transcription has been initiated, synthesis of the encoded RNA requires that RNA polymerase transcribe the entire gene and not terminate prematurely. As well, the initial **primary transcripts** produced from eukaryotic genes undergo various processing reactions to yield corresponding functional RNAs. These mature, nuclear RNAs are then actively transported to the cytoplasm, as components of ribonucleoproteins.

Opportunities for additional levels of gene control beyond the regulation of transcription initiation exist in the multiple steps in the production of RNAs. In the case of a protein-coding gene, the amount of protein expressed can be regulated by controlling the stability of the corresponding mRNA in the cytosol and the rate of its translation. In addition, the cellular locations of some mRNAs are regulated, so that newly synthesized protein is concentrated where it is needed. All of the regulatory mechanisms that control gene expression following transcription initiation are referred to as *post-transcriptional control*. This chapter highlights the various steps in the synthesis of mRNA, tRNA, and rRNA following transcription initiation and the mechanisms by which RNAs and proteins are transported in and out of the nucleus in eukaryotic cells. Relevant examples of how regulation of these steps contributes to the control of gene expression is also presented.

PART B: *Reviewing Concepts*

11.1 *Transcription Termination*
(lecture date _____)

1. Describe how the concentration of tryptophan influences transcription of the *E. coli trp* operon. Compare this regulation with the regulation of the *lac* operon.

2. Antitermination and attenuation are two mechanisms for regulating transcription termination. What is the effect of each of these control mechanisms on elongation of nascent RNA transcripts? Name examples of transcription units regulated by each mechanism.

3. In *E. coli*, transcription-termination sites are located downstream of operons but not between the genes within a single operon. What are the consequences of this arrangement for regulation of transcription?

11.2 *Processing of Eukaryotic mRNA*
(lecture date _____)

4. You have been asked to isolate mRNA as a class from cultured carrot cells. You know that mRNA constitutes at most a few percent of the total cellular RNA and is heterogeneous in size. Although this isolation appears to be a difficult experimental

problem, you are confident that it can be done readily. Describe your approach.

5. Splicing in pre-mRNAs proceeds via two transesterification reactions. Draw a sketch showing the atoms involved in these reactions and the products formed.

6. The spliceosomal splicing cycle involves ordered interactions among a pre-mRNA and several U snRNPs. According to the current model of spliceosomal splicing, which intermediate(s) in the splicing of a pre-mRNA containing one intron should be immunoprecipitated by anti-U5 snRNP? Which additional intermediate(s) should be immunoprecipitated by anti-U1 snRNP?

7. RNAs that are capable of self-splicing have been referred to as *catalytic* RNAs. Discuss how these RNAs are similar to and different from protein enzymes.

8. What do the similarities between self-splicing group II introns and snRNAs suggest about the evolution and function of snRNAs in spliceosomal splicing?

11.3 *Regulation of mRNA Processing*
(lecture date _____)

9. Differential RNA processing plays an important role in sex determination in *Drosophila*. To a large extent this is due to differential RNA splicing as a consequence of binding of Sex-lethal (Sxl) protein to newly synthesized pre-mRNAs. In *Drosophila,* maleness might be thought of as the default developmental pathway, the pathway taken in the absence of functional Sxl protein. To what extent is this a fair summary of sex determination in *Drosophila*?

11.4 *Signal-Mediated Transport through Nuclear Pore Complexes*
(lecture date _____)

10. What observations led to the directional model of mRNP transport through the nuclear pore complex?

11. What are nuclear localization signals and why are they needed?

11.5 *Other Mechanisms of Post-Transcriptional Control* (lecture date _____)

12. What effect does RNA editing have on the tissue-specific production of apolipoprotein B?

13. Human and rodent mRNAs on the whole are metabolically fairly stable. The average mRNA half-life is 10 h. Despite this general stability, considerable variation in mRNA half-life is observed. For example, histone mRNA has a half-life of less than 30 min, which is much less than that of the typical eukaryotic mRNA. What might be the advantages to organisms of such large differences in mRNA half-life?

11.6 *Processing of rRNA and tRNA*
(lecture date _____)

14. The 28S, 18S, and 5.8S RNAs in eukaryotic ribosomes are present in equimolar amounts. How is this result assured?

15. What are the main features of splicing in pre-tRNA that distinguish it from splicing in pre-mRNA?

PART C: *Analyzing Experiments*

11.1 *Transcription Termination*

16. (•) The expression of the *trp* operon in *E. coli* is controlled by a number of genetic elements including an attenuator, an operator, a promoter, and a repressor gene. Describe how you would distinguish biochemically the phenotype of mutations in the following:

a. an attentuator

b. an operator

c. a promoter

d. a repressor gene

11.2 *Processisg of Eukaryotic mRNA*

17. (•) In the R-loop technique, RNA is hybridized to DNA molecules and the base-paired complexes then are visualized in the electron microscope. When this technique is performed with mRNA and genomic DNA, introns appear as loops of double-stranded intervening DNA projecting from RNA-DNA hybrid regions. R-loop analysis of fetal globin mRNA hybridized to genomic DNA indicates the presence of one intron, but a complete sequence comparison of the cDNA and genomic DNA indicates the presence of two introns.

a. Draw a sketch illustrating the general appearance of electron micrographs of fetal globin mRNA hybridized to a genomic DNA fragment containing the globin gene.

b. How do you reconcile the conclusions reached based on each method?

c. Why is the sequence comparison made between cDNA and genomic DNA rather than directly between mRNA and genomic DNA?

11.3 *Regulation of mRNA Processing*

18. (••) Transcription of protein-coding genes by eukaryotic RNA polymerase II is a complex process in which termination may occur at any one of a number of sites downstream from the last exon. Early sequencing studies of cDNA indicated that the sequence AAUAAA or in occasional cases AUUAA was located 10–35 nucleotides upstream of the poly-A addition site. Transcription of genes in which this sequence is mutated yields a pre-mRNA that does not undergo polyadenylation and is rapidly degraded. You have inserted an AAUAAA sequence into a human sialyltransferase gene 250 nucleotides upstream from the normal site.

a. What effect should this insertion have upon the length of the primary transcript?

b. What effect should this insertion have upon the length of the resulting mRNA?

c. What effect should this insertion have upon the length of the resulting protein?

d. What effect should this insertion have upon the function of the resulting protein?

11.4 *Signal-Mediated Transport through Nuclear Pore Complexes*

19. (••) Autoradiography of pulse-labeled cells can identify the sites of biosynthetic activity and product accumulation in cells. Would autoradiographic grains be localized over the nucleus or cytoplasm in eukaryotic cells subjected to the following treatments? Why?

a. 5-min [³H]uridine pulse

b. 5-min [³H]thymidine pulse

c. 2-h [³H]uridine pulse

d. 2-h [³H]thymidine pulse

e. 5-min [³H]uridine pulse followed by a 2-h chase in precursor-free media

f. 5-min [³H]thymidine pulse followed by a 2-h chase in precursor-free media

20. (•••) Some of the proteins involved in splicing of pre-mRNA have been shown to contain the SR motif, which has the sequence S/T-P-X-R/K (Ser/Tyr-Pro-X-Arg/Lys) where X is any amino acid residue; this motif can be phosphorylated at the S/T residue. Many of these SR proteins are associated with snRNPs. SR proteins are concentrated in the nucleus of interphase cells in what appears by light microscopy as "speckles." These speckles are believed to act as storage sites for SR proteins, other splicing proteins (e.g., Sm), and associated snRNAs. Nuclear speckles break down and reform as cells progress through mitosis.

Because phosphorylation is known to regulate the subcellular localization of various nuclear proteins (e.g., lamins) and the activity of numerous proteins, a reasonable hypothesis is that phosphorylation controls the localization and activity of SR proteins. To study this hypothesis, Gui and Fu at the University of California, San Diego together with Lane at Harvard University sought to identify a protein kinase that phosphorylates SR proteins. In their initial experiments, they applied a mitotic extract prepared from HeLa cells, a human cell line, to a phosphocellulose column and eluted the column with increasing concentrations of NaCl. The resulting fractions were then assayed for the ability to phosphorylate H1 histone or an SR protein in the presence of [γ-³²P]ATP. The resulting autoradiograms are diagrammed in Figure 11-1.

a. Are H1 histones and SR proteins phosphorylated by the same protein kinase?

b. Why use a phosphocellulose column in this experiment?

c. How would you determine if SR protein kinase had any effect on nuclear speckles and if this effect was specific for SR proteins only?

d. How would you investigate if the activity of SR protein kinase was regulated during the cell cycle (i.e., as cells progress from one mitosis to the next)?

e. How would you determine if the SR protein kinase were a member of a known protein family?

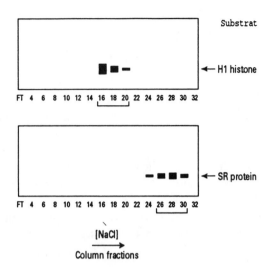

Figure 11-1

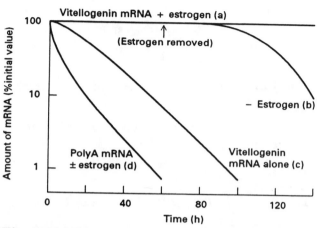

Figure 11-2

11.5 *Other Mechanisms of Post-Transcriptional Control*

21. (•••) One level at which gene expression theoretically can be controlled in eukaryotic cells is the stability of cytoplasmic mRNA. Shapiro and colleagues investigated the balance between the rate of synthesis and degradation of vitellogenin mRNA in a *Xenopus laevis* liver-cell culture system. They found that addition of estrogen, a steroid hormone, to the culture greatly enhanced the synthesis of vitellogenin mRNA and also affected the stability of mRNA in *Xenopus* liver cells. The results of assays for vitellogenin mRNA and poly A–hr containing RNA over time are shown in Figure 11-2.

   a. What assays could be used to obtain the data in Figure 11-2?

   b. What is the effect of estrogen on the stability of vitellogenin mRNA in *Xenopus* liver cells?

   c. Is the estrogen effect mRNA specific?

   d. Is the estrogen effect reversible? What does this suggest about the nature of the associations that mediate this effect?

11.6 *Processing of rRNA and tRNA*

22. Figure 11-3 shows the results of a experiment in which cultured HeLa cells were labeled with [³H]uridine. At the indicated times, whole-cell extracts were subjected to rate-zonal centrifugation and the fractions were assayed for radioactivity (black solid curves) and UV absorption (gray curves).

   a. Which cellular component is represented by the radioactivity curves and which by the UV absorption curves?

   b. Explain why the sedimentation profile of [³H]uridine incorporation changes over time, whereas the distribution of UV absorption is constant.

   c. What molecular species corresponds to the 45S peak in radioactivity in panel B of Figure 12-3? Explain the shift in the radioactivity profile (black curve) that occurs between the 15- and 60-min labeling periods.

   d. Explain why addition of actinomycin D (dashed curve) at 15 min leads to elimination of the labeled 45S and 32S peaks in panel C.

23. Human 5.8S, 18S, and 28S rRNA are, respectively, 160 bases, 1.9 kb, and 5.1 kb in length. For many naturally occurring RNA molecules, the sedimentation coefficient of the molecule, the S value, can be related to molecular weight (MW) by the equation $S = (constant) (MW)^{1/2}$.

   a. Calculate the length in kilobases of 45S pre-rRNA assuming that the ratio between the

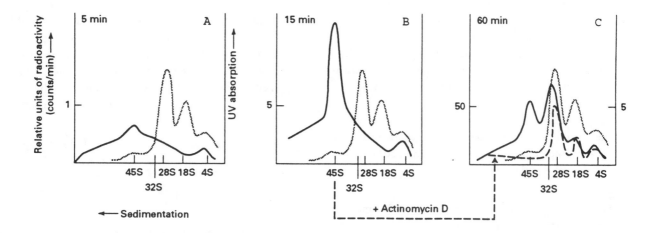

Figure 11-3

number of bases and molecular weight is the same for 28S rRNA and 45S pre-rRNA.

b. Calculate the percentage of the 45S pre-rRNA molecule that is present in each of the rRNAs derived from it.

c. What percentage of a 45S pre-rRNA molecule is metabolically stable and what percentage is rapidly degraded in the nucleus?

Answers

1. As with the *lac* operon, transcription of the *trp* operon is regulated by a small molecule. However, lactose interacts with the *lac* repressor to induce the *lac* operon, while tryptophan interacts with the *trp* repressor to prevent transcription of the *trp* operon. A second mechanism, attenuation, also regulates the *trp* operon. An attenuator site is located in the leader sequence at the 5' end of the polycistronic *trp* mRNA. This sequence can form alternative stem-loop structures that regulate the progression of RNA polymerase past the leader. When the intracellular concentration of tryptophan is high, the 5' region 1 of the attenuator site, which contains two Trp codons, is rapidly translated; the remainder of the leader then can form a stem-loop structure followed by a series of U residues, leading to ρ-independent termination of transcription. When tryptophan is scarce, a ribo-

some pauses over region 1 of the attenuator, due to lack of charged tRNA^{Trp}, allowing an alternative stem-loop structure to form. Because this stem loop is not followed by a series of U residues, RNA polymerase progresses beyond the leader and the entire *trp* operon is transcribed.

For further study, see text section "Premature Termination by Attenuation Helps Regulate Expression of Some Bacterial Operons," p. 405 and Figure 11-2 p. 406.

2. Attenuation causes premature termination of transcript elongation, while antitermination prevents premature termination of elongation. Attenuation regulates the *E. coli* operons encoding the enzymes for biosynthesis of tryptophan, phenylalanine, and histidine, as well as some other operons. Antitermination controls transcription of the λ-phage genome and HIV gene. In order for antitermination to occur in λ-phage and HIV transcription, a virus-encoded protein—N for λ and Tat for HIV—interacts with a stem-loop structure in the nascent viral transcript, called *nut* in λ and TAR in HIV. Various host-cell proteins then interact with RNA polymerase and with N and Tat to form a large complex that acts to prevent transcription termination by looping of the growing mRNA. Thus antitermination involves stem-loop structures and RNA-binding proteins.

Attenuation of bacterial operons also involves stem-loop structures. In the case of operons encoding amino acid biosynthetic enzymes, no RNA-binding proteins are involved. This mechanism can only operate when transcription and translation are coupled, as they are in prokaryotes. However, attenuation of other *E. coli* operons (e.g., *bgl* operon) is regulated by an RNA-binding protein. In the absence of this protein, a terminating stem loop forms in the leader of the nascent mRNA, thereby preventing continued transcription. When this protein is present, it binds to the leader sequence, preventing formation of the terminating stem loop, so that transcription of the entire operon proceeds.

For further study, see text section "Pre-mature Termination by Attenuation Helps Regulate Expression of Some Bacterial Operons," p. 405; "Sequence–Specific RNA-Binding Proteins Can Regulate Termination by *E. coli* RNA Polymerase," p. 407; and Figures 11-3 and 11-5, pp. 407 and 408.

3. The presence of termination sites between operons permits them to be independently regulated. The absence of termination sites between genes within a single operon permits all the genes to be regulated coordinately.

For further study, see text section "Pre-mature Termination by Attenuation Helps Regulate Expression of Some Bacterial Operons," p. 405.

4. All eukaryotic mRNAs, with the exception of histone mRNAs, have a 3′ poly-A tail, which is about 200 nucleotides long in higher eukaryotes. The poly-A tail is added to pre-mRNA during processing in the nucleus and is not found in other RNA species. Thus mRNA can be readily separated from other RNAs by affinity chromatography with a column to which short strings of thymidylate (oligo-dT) are linked to the matrix.

For further study, see text section "Pre-mRNAs Are Cleaved at Specific 3′ Sites and Rapidly Polyadenylated," p. 413.

5. See Figure 11-4 on the following page. In the first reaction, the phosphoester bond between the 5′ G of the intron and the last (3′) nucleotide in exon 1 is exchanged for a phosphoester bond between the G and the 2′ oxygen of the A at the branch point. In the second reaction, the phosphoester bond between the 3′ G in the intron and the first (5′) nucleotide in exon 2 is exchanged for a phosphoester bond with the last nucleotide in exon 1. No energy is consumed in these reactions, and the intron is released as a circular structure called a lariat.

For further study, see text section "Splicing Occurs at Short, Conserved Sequences in Pre-mRNAs via Two Transesterification Reactions," p. 415, and Figure 11-16, p. 417.

6. Two different potential intermediates should be immunoprecipitated by anti-U5 snRNP: (1) a structure in the process of joining the two exons together but still containing the intron and (2) a structure that contains the excised intron in lariat form. Both of these structures contain U5 snRNP and other snRNPs. Anti-U1 snRNP should precipitate two additional intermediates: (1) the pre-mRNA with U1 snRNP bound to the 5′ end of the intron and (2) a structure consisting of the pre-mRNA, U1 snRNP, and U2 snRNP bound to the branch site. See MCB, Figures 12-23 and 12-24, pp. 504 and 506.

For further study, see text section "Spliceosomes, Assembled from snRNPs and a Pre-mRNA, Carry Out Splicing," p. 416, and Figures 11-17 and 11-19, pp. 418 and 419.

7. Self-splicing RNA is catalytic in the sense that it increases the rate of a reactions as protein enzymes do. However, the substrates of protein enzymes are separate molecules; the intron substrate of a self-splicing RNA is part of the RNA molecule itself. Thus one enzyme molecule carries out its characteristic reaction multiple times on different substrate molecules, whereas self-splicing RNA can cleave itself only once.

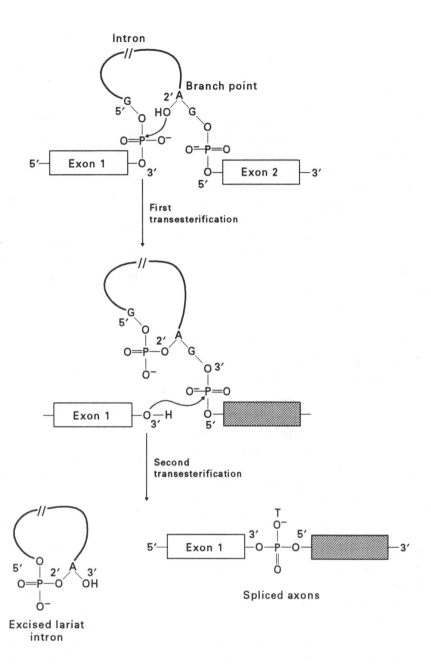

Figure 11-4

For further study, see text section "Self-Splicing Group II Introns Provide Clues to the Evolution of snRNAs," p. 419.

8. Both group II introns and the U snRNAs in spliceosomes fold into complex secondary group structures containing stem-loops. Moreover, group II self-splicing and spliceosomal splicing occurs via two transesterification reactions. These similarities suggest that ancestral group II introns evolved by losing internal sequences that became the U snRNAs, which function as trans-acting "enzymes" whose "substrates" are the introns remaining in pre-mRNA. Support for this hypothesis comes from experiments with deletion mutants of group II introns, which have lost the ability to self-splice. Addition of small RNAs that contain the deleted sequences to such mutant group II introns restores self-splicing in vitro.

For further study, see text section "Self-Splicing Group II Introns Provide Clues to the Evolution of snRNAs," p. 419.

9. In *Drosophila,* production of Sxl protein leads to a complicated set of regulated RNA splicing events in which functional Sxl protein is an active participant. Males lack functional sex-lethal protein and hence what might be thought of as a set of passive default RNA-splicing events happen in males. However, in the end, sex determination in *Drosophila* is the result of an active process of gene repression. In females the expression of a female-specific Double-sex protein represses transcription of genes required for male sexual development. In males the expression of a male-specific Double-sex protein represses transcription of genes required for female sexual development. Hence, in both *Drosophila* males and females, sexual development requires active repression steps and is not a passive default pathway.

For further study, see text section "A Cascade of Regulated RNA Splicing Controls *Drosophila* Sexual Differentiation," p. 423, and Figure 11-25, p. 424.

10. From studies on the salivary gland Balbani rings of the insect *Chironomous tentans*, investigators were able to visualize the transport of fully processed mRNAs and associated proteins (mRNPs) in the form of 50-nm coiled structures with electron microscopy. Electron micrographs of sections of salivary gland cells shows that mRNPs move through the nuclear pores to the cytosol. As they uncoil on the cytoplasmic side, they immediately become associated with ribosomes, indicating that the 5' end leads the way through the complex.

For further study, see text section "Nuclear Pore Complexes Actively Transport Macromolecules between the Nucleus and Cytoplasm," p. 427, and Figure 11-31, p. 429.

11. A nuclear-localization signal is a domain of amino acids within a protein that allows for the nuclear import of that protein. These domains are required for proteins which eventually function inside the nucleus, as all proteins are synthesized in the cytoplasm. These domains are also needed because most transport of proteins into the nucleus occurs via an active process rather than a passive one.

For further study, see text section "Receptors for Nuclear-Localization Signals Transport Proteins into the Nucleus," p. 432.

12. The *apo-B* mRNA produced in the liver has the same sequence as the exons in the primary transcript. This mRNA is translated into Apo-B100 which contains an N-terminal domain that associates with lipids and a C-terminal domain that binds to LDL receptors. In the intestine, however, the *apo-B* mRNA is edited at the CAA codon in exon 26 producing a stop codon (UAA). Translation of this mRNA results in the production of a different protein, Apo-B100, which contains only the N-terminal domain.

For further study, see text section "RNA editing Alters the Sequences of Pre-mRNAs," p. 437, and Figure 11-39, p. 437.

13. At least, two types of general explanations can be proposed for the existence of some short half-life mRNAs in higher eukaryotes. First, the protein encoded by a particular mRNA may have only a short-term physiological role. If the protein itself is only needed for a brief time, then the corresponding mRNA should be unstable, so that unnecessary protein is not produced. Second, the encoded protein may be very tightly coupled to an ordered assembly process in cells. This is the case with histone mRNAs. DNA replication is very closely coupled to the formation of new nucleosomes of which histones form the protein core. Because of the short half-life of the histone mRNAs, anything that interferes with transcription of histone genes will lead in short order to a cessation of DNA replication. There is no point in a cell replicating under conditions that fail to support transcription. Histones as proteins are relatively metabolically stable.

For further study, see text section "Stability of Cytoplasmic mRNAs Varies Widely," p. 440.

14. The 45S pre-rRNA encodes one copy of each of these mature rRNA molecules. Processing of one pre-rRNA molecule thus produces one molecule of 5.8S, 18S, and 28S rRNA, which associate with proteins to form the ribosomal subunits.

For further study, see text section "Pre-rRNA Genes Are Similar in All Eukaryotes and Function As Nucleolar Organizers," p. 443, and Figure 11-47, p. 444.

15. Splicing of pre-tRNA does not involve spliceosomes. In the first step, an endonuclease-catalyzed reaction excises the intron, which is released as a linear fragment, and a 2′,3′-cyclic monophosphate ester forms on the cleaved end of the 5′ exon. A multistep reaction that requires the energy derived from hydrolysis of one GTP and one ATP then joins the two exons. In contrast, premRNA splicing occurs in spliceosomes, involves two transesterification reactions, releases the intron as a lariat structure, and does not require GTP. Although these transesterification reactions do not require ATP hydrolysis, it probably is necessary for the rearrangements that occur in the spliceosome.

For further study, see text section "Splicing of Pre-rRNAs Differs from Other Splicing Mechanisms," p. 443, and Figure 11-53, p. 448.

16a. The biochemical phenotype of these mutations can be distinguished by comparing the amount of trp leader mRNA and full-length trp transcripts in mutant and wild-type strains. Attenuator mutations affect the incidence of transcription completion under varying tryptophan conditions but have no effect on the rate of transcription initiation. Thus the biochemical phenotype of attenuator mutants will show differences in the ratio of full-length RNA to leader RNA relative to that of wild-type.

16b. Operator mutants affect the rate of transcription initiation for the trp operon but not the responsiveness of the operon to tryptophan levels. Operator mutants will show differences in the amount of leader RNA relative to that in wild-type.

16c. Promoter mutations affect the rate of transcription initiation for the trp operon but not the responsiveness of the operon to tryptophan levels. Promoter mutants will show differences in the amount of leader RNA relative to that in wild-type.

16d. Repressor mutations affect the rate of transcription initiation for the trp operon but not the responsiveness of the operon to tryptophan levels. Operator mutants will show differences in the amount of leader RNA relative to that in wild-type.

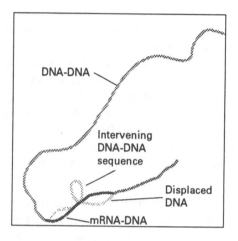

Figure 11-5

17a. See Figure 11-5. When mRNA and genomic DNA are mixed under appropriate conditions, the mRNA base-pairs with its complementary sequences in the DNA, displacing the noncomplementary strand of a double-stranded DNA molecule to generate an RNA-DNA hybrid region of double-stranded thickness and a displaced single-stranded DNA region. The occurrence of an intervening sequence (intron) within the genomic DNA will result in a double-stranded DNA bubble—the R loop—containing no sequences complementary to the mRNA.

17b. In order for an intron to be visualized by the R-loop technique, the double-stranded DNA loop (bubble) must be of a certain size. If the loop corresponding to a particular intron is not large enough to be visualized in the electron microscope, the intron will not be detected by the R-loop technique but will be detected by direct sequence comparison. In fact, fetal and adult β-like globin genes contain one long intron (850–886 bp) and one short intron (122–130 bp). The short intron is not revealed by R-loop analysis.

17c. DNA can be cloned readily and cut specifically into defined fragments by restriction endonucleases, but RNA cannot. Therefore, mRNA must be copied into cDNA for both amplification purposes (i.e., cloning) and fragmentation purposes. The resulting DNA fragments then are sequenced a few hundred nucleotides at a time. Messenger RNA is not used directly because of these problems.

18a. Transcription by RNA polymerase II terminates at any one of multiple sites 0.5-3 kb downstream from the 3′ end of the last exon in the transcript. Since there is little, if any, relationship between termination and poly-A addition signal, insertion of an additional upstream AAUAAA sequence most likely would not affect termination. Thus the primary transcript would be essentially the same length as that produced from the wild-type gene.

18b. The 3′ end of a mature mRNA is generated by cleavage and subsequent polyadenylation of the primary transcript (pre-mRNA). The AAUAAA sequence is one of two sequences that signals poly-A modification of pre-mRNA. The other is a GU- or U-rich region located ≈60–85 nucleotides downstream of the cleavage/poly-A addition site. Both signals are needed for efficient cleavage. Hence, the effect of inserting an extra AAUAAA sequence 250 nucleotides upstream from its normal site would depend on whether a GU- or U-rich region is located an appropriate distance downstream from the inserted sequence. If such a GU- or U-rich region is present, then the primary transcript produced from the altered gene could be cleaved and polyadenylated at the inserted poly-A site, generating an alternative mRNA that is approximately 250 nucleotides shorter than wild-type mRNA. If this second site is not present, then cleavage/polyadenylation could occur only at the wild-type poly-A site, generating an mRNA of normal length.

18c. Once again, there are several possibilities depending on whether a GU- or U-rich region is present downstream from the new AAUAAA sequence, and whether the last exon is greater than 250 nucleotides from the new site. If the primary transcript produced from the altered gene could be cleaved and polyadenylated at the inserted poly-A site, generating an alternative mRNA that is approximately 250 nucleotides shorter than wild-type mRNA, and if this new poly A site did not interrupt the coding reigon, then the translated protein would be the same length as the normal wild-type protein. If, however, the new poly A site interupts the coding region, the translated protein will be truncated (i.e., shorter).

18d. If the new poly A site has interrupted the coding region, then the shorter protein may not function normally if the amino acids not present in the truncated protein are important for function.

19a. Almost all the autoradiographic grains will be over the nucleus, following a brief (5-min) uridine pulse, because almost all RNA synthesis in eukaryotic cells occurs in the nucleus and a 5-min pulse is too short a time for much newly synthesized RNA to be transported from the nucleus to cytoplasm.

19b. Almost all grains will be over the nucleus, following a brief (5-min) thymidine pulse, because almost all DNA synthesis in eukaryotic cells occurs in the nucleus.

19c. Most grains will be over the cytoplasm, following a lengthy (2-h) labeling with uridine, because RNA accumulates in the cytoplasm. However, some grains will be over the nucleus because RNA synthesis continues throughout the labeling period; these nuclear grains represent newly formed RNA that has not yet moved to the cytoplasm.

19d. Almost all grains will be over the nucleus, following a lengthy (2-h) labeling with thymidine, because almost all DNA accumulation, as well as synthesis, occurs in the nucleus.

19e. Almost all grains will be over the cytoplasm, following a 5-min pulse labeling and a 2-h chase in precursor-free media, because labeled nuclear RNA formed during and after a short uridine pulse moves to the cytoplasm during a 2-h chase. What does not move to the cytoplasm is degraded in the nucleus.

19f. Almost all the grains will be over the nucleus, following a 5-min thymidine pulse and a 2-h chase in precursor-free media, because almost all the labeled DNA formed during and after a short thymidine pulse remains in the nucleus during a 2-h chase.

20a. The kinase that phosphorylates H1 histone elutes in fractions 15–21, while the one that phosphorylates SR proteins elutes in fractions 24–30 as indicated by the intensity of the phosphorylated products shown in Figure 12-7. These findings indicate that H1 histone and SR proteins are phosphorylated by different protein kinases.

20b. Since ATP (adenosine triphosphate) is a substrate for protein kinases, these enzymes would be expected to have an affinity for phosphate groups and thus would bind to a phosphocellulose column. If the affinity for phosphate groups differed from kinase to kinase, which is likely, they would be eluted at different ionic concentrations. In this experiment, the H1 kinase eluted at 0.35 M NaCl and the SR kinase at 0.6 M.

20c. One approach would be to incubate interphase cells whose plasma membrane had been rendered permeable to proteins by a mild detergent treatment with a crude cellular extract from mitotic cells, which ought to contain active SR kinase. Alternatively, fractions with SR kinase activity could be used. After incubation, the distribution of SR proteins in the nuclei could be scored using a fluorescent antibody specific for SR proteins. If SR kinase affects nuclear speckles, then the distribution of SR proteins in the treated cells should differ from that in untreated cells. Antibodies to other nuclear proteins could be used to determine if any effect of SR kinase was specific to SR proteins. As a further control, the interphase cells could be incubated with H1 histone kinase. Similar experiments could be performed with isolated nuclei rather than permeabilized cells.

20d. The simplest approach would be to determine whether extracts prepared from cells in various portions of the cell cycle exhibited different effects on the distribution of SR proteins in interphase cycles, as described in answer 69c. A second approach would be to prepare SR kinase fractions by phosphocellulose chromatography of extracts from cells in various portions of the cell cycle and assay their kinase activity with SR protein. Although the second approach involves much more work, it does have the advantage that the SR kinase is at least partially purified, so that a wide range of cytosolic proteins, which might directly or indirectly affect the distribution of SR proteins, are not present.

20e. The best way to determine if SR protein kinase is related to other known proteins and hence a member of a protein family would be to determine its amino acid sequence and compare this sequence to that of other known proteins. Several approaches are possible for obtaining the sequence of a protein. A routine starting point is to purify microgram amounts of the protein and determine the sequence of 10–20 residues at its N-terminal end by an automated Edman degradation procedure. This sequence information then is used to prepare oligonucleotide probes for screening cDNA libraries. Once a cDNA clone encoding the protein of interest is identified, it can be sequenced and the amino acid sequence of the entire protein can be deduced from the nucleotide sequence.

21a. The data in Figure 11-2 represent the amounts of vitellogenin mRNA and total mRNA (poly A-containing RNA) in the cytoplasm. After the liver cells are disrupted and nuclei removed by low-speed centrifugation, the total mRNA could be determined by affinity chromatography with a poly-T column and vitellogenin mRNA could be determined by Northern blotting with a specific DNA probe.

21b. Comparison of curve (a) and curve (c) indicates that the half-life of vitellogenin mRNA is longer in estrogen-treated than in untreated cells. Thus estrogen appears to increase the stability of vitellogenin mRNA.

21c. The estrogen effect on mRNA stability appears to be specific for vitellogenin mRNA, as the half-life of poly A-containing mRNA is the same in treated and untreated cells (curve d). However, the half-life value for total poly A-containing mRNA is an average for many different mRNA species, some of which are present in small amounts. For this reason, the presence of minor mRNA species that are stabilized by estrogen treatment cannot be proved or disproved from these data.

21d. The stabilization of vitellogenin mRNA by estrogen is reversed when estrogen is removed (curve b). This finding suggests that the stabilization of vitellogenin mRNA more likely results from reversible protein-mRNA interactions than from formation of stable covalent modifications of the mRNA.

22a. The UV absorption curves correspond to the total cellular RNA; this is largely preexisting rRNA and tRNA present in the cells before labeling. The radioactive curves correspond to newly synthesized RNA.

22b. During the longer labeling periods, some of the newly synthesized radiolabeled RNA undergoes processing, producing species with different sedimentation properties. Because the UV-absorption profile reflects the bulk, steady-state distribution of RNA among various RNA species, this profile is constant with time.

22c. The 45S peak in panel B is newly synthesized, unprocessed pre-rRNA. This pre-rRNA subsequently is cleaved, yielding 32S and 20S pre-rRNA molecules. These in turn are cleaved to produce the mature 18S and 28S rRNAs (see MCB, Figure 12-50, p. 530). However, since incorporation of label continues between 15 and 60 min, the radioactive profile in panel C reflects the presence of all these rRNA species.

22d. In this experiment, actinomycin D functions as a chase because it blocks transcription. Following addition of actinomycin D, no additional synthesis of labeled 45S pre-rRNA occurs. Between 15 and 60 min, all the previously synthesized labeled pre-rRNA is cleaved to yield 28S and 18S rRNAs.

23a. Let X = number of bases in 45S pre-rRNA and set up the following ratio: $45/28 = X^{1/2}/5100^{1/2}$. Solving for X gives 13.2 kb.

23b. The percentage of 45S pre-rRNA in 5.8 S RNA = 1.2 percent; in 18S RNA = 14 percent; and in 28S RNA = 39 percent. These values are calculated by dividing the number of bases in each species by the number of bases in pre-rRNA and multiplying the result by 100.

23c. The metabolically stable portion of pre-rRNA forms 5.8S, 18S, and 28S rRNA. Thus you can calculate the percentage that is metabolically stable from the ratio of the total number of bases in the three smaller rRNAs to the number of bases in pre-rRNA:

$$(0.16 \text{ kb} + 1.9 \text{ kb} + 5.1 \text{ kb}) \div 13.2 \text{ kb} = 0.54, \text{ or } 54 \%$$

The metabolically unstable proportion is calculated by difference: 100 − percentage stable = 100 − 54 = 46%.

12

DNA Replication, Repair, and Recombination

PART A: *Chapter Summary*

The fidelity of DNA replication and DNA repair is important for maintaining the integrity of an organism's genome. This chapter discusses the mechanisms for replication and repair of DNA and for recombination between DNA molecules. DNA replication is semiconservative, bidirectional, and initiates at specific chromosomal sites. The DNA replication machinery is a multiprotein complex, consisting of proteins with helicase, polymerase, and primase activities, that synthesizes both the leading and lagging strands concurrently. The replication machinery in eukaryotes is generally similar to that of bacteria. In eukaryotes, a specialized ribonucleoprotein complex called telomerase is needed to replicate the ends of linear chromosomes.

There are two different classes of topoisomerases, which differ in their mode of action. Type I topoisomerases can relax DNA by nicking and closing one strand of DNA. Type II topoisomerases break and rejoin double-stranded DNA molecules and play an important role in removing positive supercoils generated at the replication fork and for separating replicated circular molecules and linear chromosomes.

DNA repair mechanisms can be classified into three broad categories: mismatch repair, excision repair, and repair of double-strand breaks by an end joining process. In the case of mismatch repair, the DNA strand with the incorporated mismatch can be identified by its methylation state. Newly replicated DNA is unmethylated, whereas the parental strand is methylated.

Recombination between DNA molecules is a mechanism for generating genetic diversity. The recombination process can be modeled by the Holliday model and the double-strand break model for genetic recombination. Both models involve the nicking of DNA, strand invasion, formation of crossed-strand Holliday structures, branch migration to form heteroduplex regions, and resolution of the Holliday structures to form recombinant and non-recombinant products. The proteins that mediate recombination in *E. coli* have been identified.

PART B: *Reviewing Concepts*

12.1 *General Features of Chromosomal Replication* (lecture date _____)

1. Describe the experimental evidence for bidirectional replication of DNA.

2. Describe the evidence for the semiconservative mechanism of DNA replication based on the Meselson-Stahl experiment.

12.2 *The DNA Replication Machinery* (lecture date _____)

3. Describe the model for concurrent replication of leading and lagging DNA strands at a growing fork in *E. coli*.

4. What is the role of telomerase in the replication of eukaryotic chromosomes?

5. What parallels exist in the structure and function of the prokaryotic and eukaryotic DNA polymerases involved in DNA replication?

12.3 *The Role of Topoisomerases in DNA Replication* (lecture date _____)

6. How does the action of type I topoisomerases differ from type II topoisomerases?

7. Describe the role of DNA gyrase during replication.

12.4 *DNA Damage and Repair and Their Role in Carcinogenesis* (lecture date _____)

8. How does the *E. coli* mismatch repair machinery distinguish between the newly replicated strand and the template strand?

9. Describe how thymine-thymine dimers are repaired in *E. coli*.

10. Describe the role of the SOS repair system in bacteria.

12.5 *Recombination between Homologous DNA Sites* (lecture date _____)

11. Describe the Holliday model for genetic recombination.

12. How does the double-strand break model for genetic recombination differ from the Holliday model for genetic recombination?

13. Describe the role of the *E. coli* RecBCD and Rec A proteins in the formation of a Holliday structure.

PART C: *Analyzing Experiments*

12.1 *General Features of Chromosomal Replication*

14. (••) Shown in Figure 12-1 are chromosome homologues from an experiment designed to determine if different portions of eukaryotic chromosomes replicate in different portions of the DNA synthetic period (S phase) of the cell cycle. In this experiment, white blood cells, which will divide a few times out-

side of the body, were cultured with bromodeoxyuridine (BrdU) for 48h and then placed in a medium containing [^{3}H]-thymidine for a brief period of time. Cells were then incubated in isotope-free medium and chromosome spreads were prepared. The regions of [^{3}H]-thymidine incorporation were detected by autoradiography (black dots in Figure 12-1a). The chromosomes also were stained with a fluorescent dye that binds well to DNA containing thymidine and poorly to DNA substituted with BrdU (Figure 12-1b). The chromosome homologues shown are from the earliest post-labeling time point to show [^{3}H]-thymidine incorporation.

a. To which portions of the chromosome is [^{3}H]-thymidine incorporation restricted? Is the labeled region replicated early or late in the DNA synthetic period?

b. The chromosome staining pattern also reflects the chromosome replication pattern. Based on the staining pattern in Figure 12-1b, which portions of the chromosome are replicated in early, mid-, and late portions of the DNA synthetic period?

c. Which of the two methods—fluorescent dye staining or autoradiography following tritium incorporation—has the higher resolution?

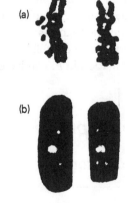

(a)

(b)

Figure 12-1

d. Based on the labeling and staining patterns, do the telomeric regions of the chromosome replicate early or late in the DNA synthetic period?

12.1 General Features of Chromosomal Replication, 12.2 The DNA Replication Machinery

15. (•••) When mammalian DNA is replicated in the presence of a protein synthesis inhibitor (e.g., emetine) and the nucleoside analog 5-bromodeoxyuridine (BrdU), the segregation of nucleosomes on newly synthesized DNA is conservative. The parental nucleosomes are associated with the leading portion of the growing fork, and no new nucleosomes are formed under these conditions. The lagging-strand DNA with no associated nucleosomes is accessible to micrococcal nuclease digestion, whereas the leading-strand DNA is protected and released as nucleosome-associated DNA; the BrdU-containing leading strand DNA can then be isolated by isopycnic centrifugation. Potentially, if strand-specific probes are available, blot hybridization of the probes with the BrdU single-stranded DNA can be used to map the origin of replication within a replicon.

The results of applying this approach to the dihydrofolate reductase region of hamster chromosomal DNA is shown in Figure 12-2. Part (a) of this figure shows the position of various (+) and (–) strand probes relative to a kilobase-pair ruler of this DNA region. Part (b) shows the blot hybridization patterns with the different probes. By convention, the (+) strand has the 5'→3' orientation (→), and the (–) strand has the opposite orientation (←).

(a) Position of probes

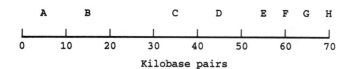

Kilobase pairs

(b) Blot hybridization patterns

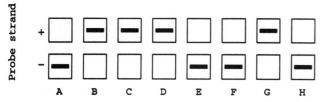

Figure 12-2

a. Indicate the direction of replication at each probe site by placing appropriate arrows on the map of the probes in Figure 12-2a. Assuming that replicons are bidirectional, how many different replicons are involved in replication of this 70- to 80-kb DNA segment? Also indicate on the probe map the position of the origin(s) of replication.

b. How do these results demonstrate that origins may *not* be centrally located within a chromosomal replicon? Propose an explanation for their noncentral location.

12.5 Recombination between Homologous DNA Sites

16. (•••) Both integration and excision of bacteriophage λ in *E. coli* requires integrase (Int), a phage-encoded protein. Int-dependent recombination does not require any high-energy cofactors or involve any degradation or synthesis of DNA. Genetic experiments suggest Int-dependent recombination proceeds via Holliday structures (chi forms), which are resolved by a cycle of nicking and ligation to yield the recombination products.

Experimentally, synthetic Holliday-structure analogs for integration and excision containing a single radioactive [^{32}P]-atom can be constructed. Two examples of such structures (A and B) are illustrated in Figure 12-3a. At the center of each of these chi forms is the "overlap" (O) region. During recombination, this region is cut at position –3/–2 (from the center) in one strand and at position +4/+5 in the other strand to generate a 7-bp overlap. Homology within the overlap region is necessary for efficient recombination.

Figure 12-3b shows the gel electrophoretic patterns, revealed by autoradiography, of the products obtained by incubating chi forms A and B with Int protein. The components of the incubation mixtures are shown at the top; lane 3 is the migration pattern of a set of standards corresponding to the segments labeled in Figure 12-3a and the unresolved chi form.

(a) ^{32}P-labeled (*) Holliday structures

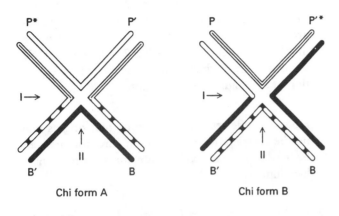

Chi form A   Chi form B

(b)

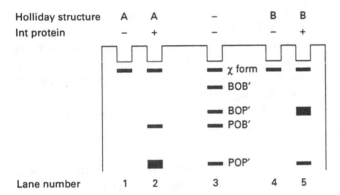

Holliday structure	A	A	–	B	B
Int protein	–	+	–	–	+

χ form
BOB'
BOP'
POB'
POP'

Lane number   1   2   3   4   5

Figure 12-3

a. Are chi forms A and B preferentially cleaved at site I or II?

b. Which recombination products from chi forms A and B are not labeled?

c. Draw a diagram showing the derivation of the radioactive and nonradioactive recombination products resulting from cleavage at site I and site II in chi form A and B.

12.2 *The DNA Replication Machinery*

17. (••) The production of a cloned sheep (Dolly) by nuclear transfer technology was a major technological breakthrough. Dolly was produced by the transfer of a nucleus from a mammary cell of a 6-year old ewe to an enucleated oocyte. The question was raised as to whether Dolly would show premature ageing as a result of "starting" with a six-year old nucleus. One hypothesis for aging proposes that telomeres shorten as cells age. Thus, P. Shiels and coworkers measured the mean size of the terminal telomere fragments for Dolly and other sheep cloned from embryonic or fetal fibroblasts.

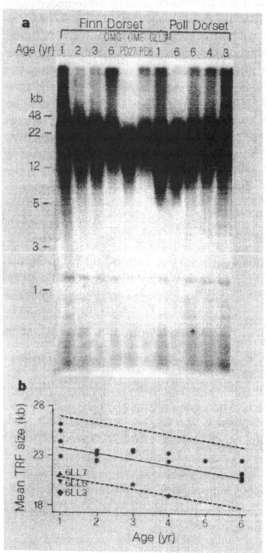

Figure 12-4

Genomic DNA was extracted from tissues of Finn Dorset and Poll Dorset sheep at different ages and then digested with a restriction enzyme. The OMG-6 (ovine mammary gland) lane represents tissue used to provide donor cells for nuclear transfer to produce Dolly. The OME lanes are primary cells derived from the OMG-6 tissues and were passaged for 27 (PD27) or 6 (PD6) population doublings in culture. Lane GLL3 represents

DNA from Dolly. The Southern blot was hybridized with a telomere-specific probe. The mean length of the terminal restriction fragment (TRF) is shown in the blot in part (a) and graphically in part (b). In the graph in part b the dots represent control sheep. The mean TRFs at one year of age were determined for Dolly (6LL3) and sheep cloned from day 9 embryos (6LL6) and day 25 fetal fibroblasts (6LL7).

a. What do you conclude about the mean size of the TRF with increasing age in control animals?

b. What effect does time in culture have on the mean size of TRFs?

c. How does the TRF for the sheep cloned by nuclear transfer compare with control sheep at one year of age?

d. Based on the available data, how might you minimize telomere shortening in animals cloned by nuclear transfer?

Answers

1. The two experimental approaches that show that DNA replicates bidirectionally involve fiber autoradiography and electron microscopy. In fiber autoradiography replicating mammalian cells are exposed first to high and then to low concentrations of [^{3}H] thymidine. The resulting fiber autoradiographic pattern showed two growing forks (heavy labeled-light labeled fibers) moving away from a central region consistent with bidirectional replication was observed. Using electron microscopy, replication "bubbles" or "eyes" were observed in SV40 DNA. The centers of these bubbles remained a constant distance from a cut end of the DNA molecule. Both of these sets of data are consistent with bidirectional replication.

    For further study, see text, pp. 455-456 and Figures 12-3 and 12-4.

2. The Meselson-Stahl experiment provided definite evidence for a semiconservative mechanism for DNA replication. *E. coli* cells were grown in a medium containing ammonium salts labeled with "heavy" nitrogen [^{15}N]. Cells were then shifted to a medium containing "light" nitrogen [^{14}N]. Samples were removed periodically and the DNA in each sample was analyzed by density gradient equilibrium centrifugation, which can separate heavy-heavy, heavy-light, and light-light DNA duplexes. The banding patterns observed were consistent with semiconservative replication, i.e., heavy-light DNA.

    For further study, see Classic Experiment 12.1 on the MCB CD-ROM, as well as text, p. 454 and Figure 12-1.

3. A single DnaB helicase moves along the lagging strand template towards its 3' end, separating the DNA duplex at the replication fork. One core polymerase adds nucleotides to the 3' end of the leading strand, while bound to the DNA by its β-subunit clamp. A second core polymerase, also bound by its β-subunit clamp, synthesizes the lagging strand as Okazaki fragments. The two core polymerases are bound together by the τ protein.

    For further study, see Animation 12-3, "Coordination of leading and lagging strand synthesis" on the MCB 4.0 CD-ROM, as well as text, pp. 463-464 and Figure 12-11.

4. Telomerase is a ribonucleoprotein complex that plays an essential role in the replication of the ends of linear chromosomes. Because all known DNA polymerases elongate DNA chains from the 3' end and require a RNA or DNA primer, the lagging strand would progressively shorten with each round of replication. Telomerase contains a reverse transcriptase activity that synthesizes DNA from an RNA template. In this way the lagging strand can be replicated in its entirety.

    For further study, see text, pp. 466–467 and Figure 12-13.

5. Both eukaryotic and prokaryotic DNA polymerases are thought to exist as dimers. Prokaryotic polymerase III is the structural and functional equivalent of a complex of eukaryotic polymerases α and δ. Polymerase III, together with the primosome, can carry out leading- and lagging-strand synthesis. In the eukaryotic complex, polymerase α plus primase synthesizes the lagging strand, and polymerase δ plus the associated PCNA synthesizes the leading strand.

    For further study, see text, pp. 464-466 and Figure 12-12.

6. Type I topoisomerases relax DNA by nicking and closing one strand of duplex DNA. Type II topoisomerases change DNA topology by breaking and rejoining double-stranded DNA.

For further study, see text, pp. 468-470 and Figures 12-14 and 12-16.

7. During the unwinding of DNA at the replication fork, positive supercoils are introduced into the duplex DNA ahead of the growing fork. These supercoils must be removed for DNA synthesis to proceed. DNA gyrase is a type II topoisomerase that removes the positive supercoils by cutting both strands of the double helix, passing another portion of the duplex through the cut, and resealing the duplex.

For further study, see text, pp. 469–470 and Figure 12-18.

8. The *E. coli* mismatch repair system recognizes and repairs single base mispairs and small insertions and deletions. The repair system can distinguish between newly replicated and template strands based on their methylation state. *E. coli* DNA contains methylated adenine residues at GATC sequences. During DNA replication, the newly synthesized strand (daughter strand) contains unmethylated adenines and the template strand (mother strand) contains methylated adenines. The *E. coli* protein *MutH* binds to this hemimethylated DNA and can distinguish the newly replicated strand from the template strand.

For further study, see text, pp. 476-477 and Figure 12-24.

9. In *E. coli* the UvrA-UvrB complex binds to DNA and scans along the DNA until it encounters a lesion, such as a thymine-thymine dimer. In an ATP-dependent reaction, the DNA at the site of the lesion is kinked. UvrA then dissociates from the complex and UvrC binds and cuts the DNA strand with the lesion. The lesion is removed and the gap is filled in by DNA polymerase and DNA ligase.

For further study, see text, p. 478 and Figure 12-26.

10. The SOS repair system of *E. coli* is induced in response to extensive DNA damage. This system is error prone because it introduces many errors in the DNA as it repairs lesions. The SOS system is induced as a last resort when error-free mechanisms are overwhelmed.

For further study, see text, p. 481.

11. The Holliday model for genetic recombination first involves the aligning of two homologous DNA molecules. A nick is introduced into one strand of each DNA molecule and the nicked strands then base pair with the complementary strand of the other DNA molecule in a process called strand invasion. The crossed-strand Holliday structure can then migrate, forming heteroduplex molecules. The DNA strands are then cut and religated to form heteroduplexes that are recombinant or non-recombinant molecules.

For further study, see text, pp. 482–483 and Figure 12-29.

12. In the double-strand break model for genetic recombination, a double strand break is introduced into one of the homologous DNA molecules. This break is then widened into gaps by a $5' \rightarrow 3'$ exonuclease. After strand invasion, the 3' end of one strand is extended by DNA polymerase to form a heteroduplex region flanked by two Holliday structures. In the Holliday model, a nick is introduced into only one strand of each DNA molecule. A single Holliday structure is formed after strand invasion. Also, there is no requirement for exonuclease and DNA polymerase activity.

For further study, see text, pp. 484–486 and Figures 12-29 and 12-31.

13. The RecBCD complex binds to free DNA ends that result form a double-strand break. The complex unwinds DNA using its helicase activity and simultaneously degrades both strands of DNA using its $5' \rightarrow 3'$ and $3' \rightarrow 5'$ exonuclease activities. When the complex encounters a CHI site, the $3' \rightarrow 5'$ exonuclease activity is inhibited and the $5' \rightarrow 3'$ exonuclease activity is enhanced, resulting in a single-stranded 3' end. The 3' end is coated with RecA protein, which then catalyzes strand invasion and formation of a Holliday structure.

For further study, see text, p. 486 and Figure 12-32.

14a. The silver grains indicating [^{3}H]-thymidine incorporation are concentrated towards the centromere.

Because these are the earliest labeled mitotic chromosomes, the labeled region must have been replicated during the late portion of the DNA synthetic period.

14b. The staining pattern indicates that two regions of the chromosome replicated late in the DNA synthetic period. The first and major region is the centromere, which shows as a brightly staining region, corresponding to the region labeled by [³H]-thymidine. The second and less bright region is about midway on each of the chromatids. In this experiment, only newly synthesized DNA that contains thymine instead of BrdU is stained.

14c. The dye staining method, which can resolve two late-replicating regions, appears to have a higher resolution than the autoradiographic method. The grains in the autoradiograph are more scattered and some are distal from the chromosome, so that only one late-replicating region (the centromere) is defined.

14d. The telomeric regions of the chromosome are certainly not late replicating, because neither silver grains nor dye is present near the ends of the chromosome. Thus the telomeres must be replicated either during the middle or early portion of the DNA synthetic period.

15a. See Figure 12-5. In order for hybridization to occur, the probe and the isolated leading-strand DNA must be complementary (i.e., have opposite 5′→3′ orientations). Because the orientation of the probe strands is known, the direction of replication at each of the probe sites can be deduced. For example, because the (−) strand of probe A hybridizes, this region of the leading strand must have the (+) orientation (→). At the position between two replicons, the growing forks of the adjacent replicons are moving towards each other (← →). Hence three replicons are involved in replication of this DNA segment (see Figure 12-5). Replicon I on the left in the figure must extend further in the left-hand direction, as only the rightward moving segment of the bidirectional fork is revealed in this experiment. Also, replicon III on the right, revealed by probes G and H, may extend further rightward; the right-hand end of this bidirectional fork is not delineated by these data.

At an origin of replication, the two growing forks within a replicon are moving away from each other

(← →). In this segment of DNA only the origins of the middle and right-hand replicons (II and III) can be mapped (see Figure 12-5). Of course, use of additional probes would locate these origins more precisely.

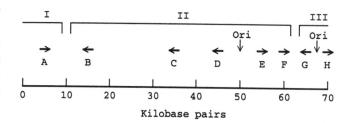

Figure 12-5

15b. The origin of replicon II clearly is to the right of the center of the replicon. This noncentral position presumably can occur because replicons are bounded by termination sequences, whose position may differ at the two ends. Although the origin of replicon III appears to be in the center, the right-hand terminus of this replicon is not mapped in this experiment. Thus this replicon also might have a noncentral origin.

16a. Chi form A is preferentially resolved at site I, as lane 2 shows more radioactivity in the POP′ band than in the POB′ band. The presence of the chi band in lane 2 indicates that some of the intermediate molecules have not been resolved. Chi form B is preferentially resolved at site II, as lane 5 shows more radioactivity in the BOP′ band than in the POP′ band. Again some of the chi B molecules have not been resolved as indicated by the presence of the chi band.

16b. The nonradioactive products from chi form A are BOB′ and BOP′. The nonradioactive products from chi form B are BOB′ and POP′.

16c. Derivation of all four products from chi form A and B is shown in Figure 12-6.

17a. From the graph in Figure 12-4b, the mean TRF for control sheep decreases with increasing age at a rate of 0.59 kilobases per year. Therefore, telomere length progressively decreases with age.

17b. From the Southern blot in part a, the mean size of the TRFs decrease with increasing time in culture (compare lanes PD27 with PD 6).

17c. The mean TRF sizes were smaller in all three nuclear transfer (cloned) sheep compared to their age matched controls. The TRF for 6LL3 (Dolly) was the smallest. In fact, the TRF for Dolly is more rep resentative of a TRF from six-year old mammary tissue, from which the donor nucleus was obtained.

17d. Based on the available data, time in culture and the source of the donor cells are important for minimizing the extent of shortening of the TRFs. Fetal or embryonic cells cultured for a minimal amount of time would likely yield the least reduction in TRF sizes of animals cloned by nuclear transfer.

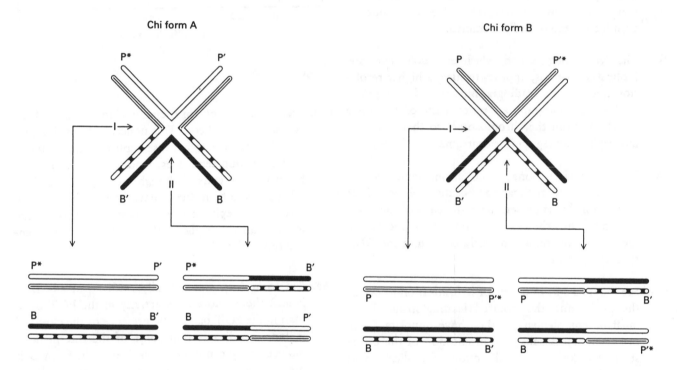

Figure 12-6

13

Regulation of the Eukaryotic Cell Cycle

PART A: *Chapter Summary*

The cell division cycle occurs in all eukaryotic cells and is characterized by four phases. These are G_1, S phase, when DNA synthesis occurs, G_2, and M phase (subdivided into prophase, metaphase, anaphase, and telophase), during which the chromosomes are segregated to daughter cells. Regulation of the cycle is effected by heterodimeric protein kinases, consisting of a regulatory subunit called a cyclin plus a catalytic subunit, a cyclin-dependent kinase (Cdk); dimers of this type are involved in *all* phases of the cell cycle. Their activity is influenced by the presence or absence of inhibitors, phosphorylation/dephosphorylation, and the polyubiquitin-targeted degradation of specific proteins by proteasomes. Other cell constituents, such as the lamins of the nuclear envelope, condensin, and myosin light chain also undergo alterations.

The logic underlying the cycle is that each regulatory event activates one phase of the cell cycle and prepares the cell for the next phase. Degradation of particular molecules ensures that the cycle is unidirectional and irreversible. The critical stages during which the cell cycle is regulated are entry into S and and entry into M phase. Checkpoints at these and other times ensure that the chromosomes are properly replicated and that previous stages have been completed before the cycle continues.

The dissection of the cell cycle has involved the use of both biochemical and molecular biological techniques. A variety of organisms have been the subject of investigation, each a model for a particular stage of the cell cycle. Frogs (*Xenopus*) and sea urchins were used to characterize biochemically the metaphase/mitosis-promoting factor (MPF), whose activity correlates with the level of cyclin B. Mutations in the yeast *S. pombe* were used to study the regulation of the kinase portion of Cdks and to show the homology between yeast and *Xenopus* MPF. Another yeast, *S. cerevisiae*, was used to investigate the mechanism of entry into S phase.

Understanding the details of the cell cycle will require further investigations, in spite of the remarkable progress made in the past two decades. These investigations are partially fueled by the intimate link between the cell cycle and the development of cancer. Many cancers are mutated in genes that regulate the cell cycle, and more information about the proteins encoded by these genes should help design therapeutic strategies.

PART B: *Reviewing Concepts*

13.1 *Overview of the Cell Cycle and Its Control* (lecture date _____)

1. Two critical regulatory mechanisms for the cell cycle are phosphorylation/dephosphorylation and degradation. How do these mechanisms apply to the anaphase-promoting complex?

2. Explain how yeast *cdc* mutants have been used to isolate the human homologs of yeast genes that regulate the cell cycle.

13.2 *Biochemical Studies with Oocytes, Eggs, and Early Embryos*
(lecture date _____)

3. What features of early *Xenopus* development made this an excellent model system for studying MPF?

4. When cycling *Xenopus* egg extracts are treated with RNase, oscillations in MPF activity are abolished. Addition of mRNA for what protein, other than cyclin B, would restore cyclical MPF activity?

13.3 *Genetic Studies with S. pombe*
(lecture date _____)

5. How does the three-dimensional structure of human Cdk2, a homolog of cdc2, inform our understanding of the regulation of the activity of this protein?

13.4 *Molecular Mechanisms for Regulating Mitotic Events* (lecture date _____)

6. Dephosphorylation is an important event that affects cellular structures during mitosis. Describe two of these events.

13.5 *Genetic Studies with S. cerevisiae*
(lecture date _____)

7. How is DNA replication regulated so that it occurs only once during the cell cycle, at START?

13.6 *Cell-Cycle Control in Mammalian Cells* (lecture date _____)

8. What is the significance of G_0 to cells within an organism as opposed to tissue culture?

13.7 *Checkpoints in Cell-Cycle Regulation*
(lecture date _____)

9. At which points in the cell cycle do checkpoint controls operate and how do they prevent progression?

Part C: *Analyzing Experiments*

13.1 *Overview of the Cell Cycle and Its Control*

10. (•)One way to assess the distribution of cells between various portions of the cell cycle is to know the amount of DNA per cell within a population. Flow cytometry with a fluorescence-activated cell sorter (FACS) can quantitate the fluorescence of individual cells within a cell population. By use of certain dyes that are nonfluorescent in solution but strongly fluorescent when bound to DNA, per cell fluorescence can be directly equated with DNA content. One such dye is Hoechst 33258, a bisbenzimide compound. These dyes are permeable to cellular membranes and very specific in their binding properties; they intercalate into the DNA double helix.

a. Figure 13-1 is a plot of cell number versus Hoechst 33258 fluorescence per cell for an exponentially growing HeLa cell population. Based on this plot, what is the relative distribution of cells in this population in the various phases of the cell cycle?

b. Sketch the expected FACS plots if these cells were arrested in each of the cell-cycle phases, namely, G_0, G_1, S, G_2, and M.

c. Is a FACS assay capable of distinguishing between all the phases of the cell cycle?

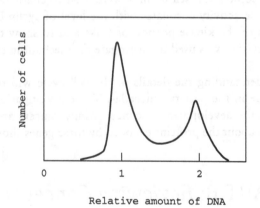

Figure 13-1

11. (•)A rapidly growing culture of mammalian cells was incubated with [³H]thymidine for 30 min, after which the radioactive thymidine was removed. At various times thereafter, cell samples were removed from the culture and subjected to autoradiography. The proportion of mitotic cells (i.e., those in metaphase) that were radiolabeled was determined from the autoradiograms and plotted as a function of time after the radioactive pulse, as shown in Figure 13.2. Note that the earliest samples have no labeled mitotic cells and that the proportion of labeled mitotic figures increases until it is nearly 100 percent in peak 1. In these cells, the duration of the M phase is 1 h.

a. What do peaks 1 and 2 in Figure 13-2 represent?

b. Are these data from a synchronized cell population? Explain.

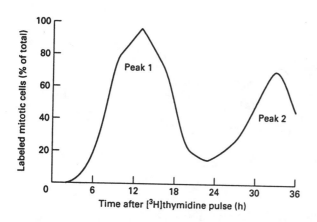

Figure 13-2

13.2 *Biochemical Studies with Oocytes, Eggs, and Early Embryos*

12. (•••)Passage of cells through mitosis releases a block to re-replication of the DNA. Experiments by Blow and Laskey with *Xenopus* extracts have provided important data regarding this phenomenon. In these experiments, intact sperm were added to a cytosolic extract from frog eggs arrested in interphase. After the DNA replicated, MPF was added (experiment 1), the replicated nuclei were isolated and added to fresh extract (experiment 2), or the replicated nuclei were isolated and permeabilized and then added to fresh extract (experiment 3). The experimental protocol and results are depicted in Figure 13-3. Based on these and other similar experiments, Blow and Laskey advanced the hypothesis that a consumable *licensing factor* controls the ability of the sperm nuclei to replicate DNA in the frog extracts.

a. In these experiments, is the licensing-factor required for the second round of replication present in the newly replicated sperm nuclei or in the frog extract? Explain your answer.

b. Propose a model consistent with these experiments for licensing-factor control of replication.

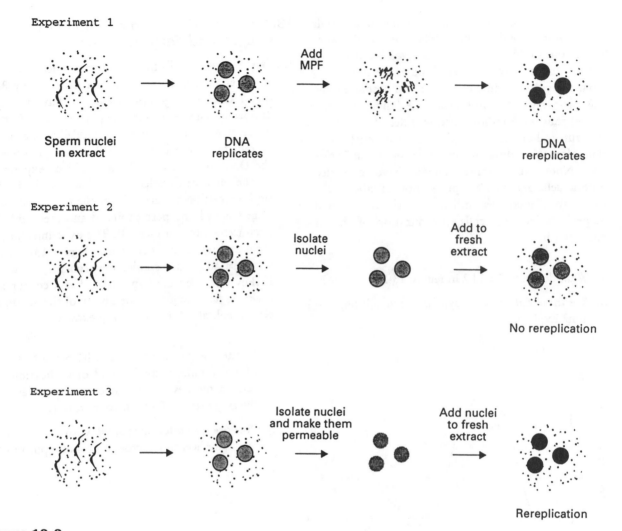

Figure 13-3

13.3 *Genetic Studies with S. pombe*

13. (•)Figure 13-4a shows various reactions involving cyclin B and Cdc2 based on mathematical modeling of cell-cycle progression in *S. pombe*. Figure 13-4b shows the cyclin levels (dashed curves) and MPF activity (solid curves) in wild-type cells and in three different deletion mutants.

a. In Figure 13-4a, indicate which molecular form has MPF activity and fill in the names of the enzymes that catalyze each of the steps indicated by the solid arrows.

b. In Figure 13-4b, indicate which genes are deleted to give the phenotypes for the panels other than wild-type. Explain your answer.

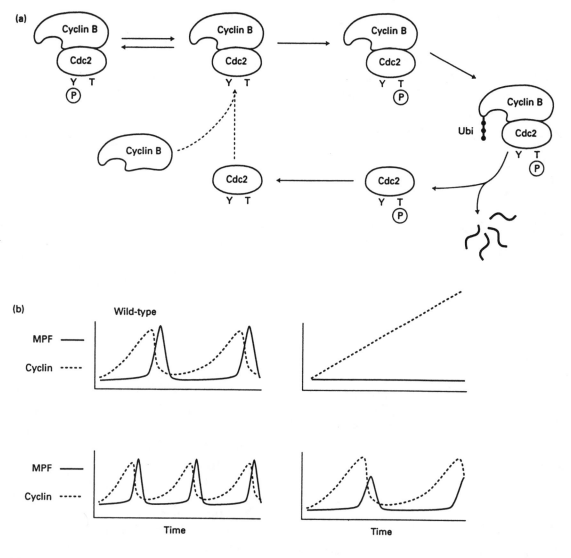

Figure 13-4

14. (●●)A number of different cell-cycle mutations including *wee1* ᵗˢ have been isolated in the fission yeast
S. pombe. Cells carrying the *wee1* ᵗˢ mutation grow
normally at the permissive temperature. However,
at the restrictive temperature, the minimum size required to enter mitosis is decreased greatly; as a
result, diminutive daughter cells are produced.

   a. If *wee1*ᵗˢ mutants are shifted from the permissive
   to nonpermissive temperature, will the first cell
   cycle completed following the shift be longer or
   shorter than normal? The duration of which portion of this first cell cycle is altered?

   b. In subsequent cell cycles at the nonpermissive
   temperature, what is the expected duration of
   the cell cycle relative to normal? What is the expected relative duration of the G_1 and G_2 phases?

13.4 *Molecular Mechanisms for Regulating Mitotic Events*

15. (●)Nocodazole is an inhibitor of microtubule polymerization whose effect is reversible. This drug often is used to synchronize mammalian cells for cell-
cycle studies. Describe how you would produce a
synchronized cell culture with this drug.

13.5 *Genetic Studies with S. cerevisiae*

16. (••)You have mutagenized a culture of the budding yeast *S. cerevisiae* and have isolated three different clones (A, B, and C). Two of the clones (A and B) exhibit temperature-sensitive growth. After shifting the cultures to an elevated temperature, you examine cells from each under a light microscope. Their appearance is depicted in Figure 13-5. Which of these clones is a wild-type, a *cdc* mutant, and a non-*cdc* mutant?

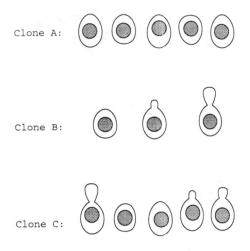

Figure 13-5

13.6 *Cell-Cycle Control in Mammalian Cells*

17. (••)Mammalian cells cultured in the absence of growth factors are arrested in the G_0 phase of the cell cycle. You are interested in dissecting the molecular requirements for relieving this growth arrest. You review the literature and find several reports quantifying the expression of new mRNAs following addition of serum, a source of growth factors, to G_0-arrested cells. Two classes of mRNA are produced, as illustrated in Figure 13-6. The translation of mRNA encoded by early-response genes is necessary for expression of the delayed-response genes. From other reports you learn that cycloheximide, an inhibitor of protein synthesis, not only blocks transcription of the delayed-response genes but also prevents the shutdown of transcription of the early-response genes that normally occurs.

a. Based on the data in Figure 13-6, is expression of the delayed-response genes required for the shutdown of the early-response genes?

b. Propose a hypothesis concerning the mechanism by which transcription of the early-response genes is shut down.

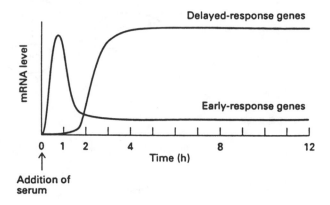

Figure 13-6

13.7 *Checkpoints in Cell-Cycle Regulation*

18. (•••)In normal eukaryotic cells, onset of anaphase, triggered by MPF inactivation resulting from cyclin B degradation, does not occur until spindle assembly is complete. Treatment with microtubule-depolymerizing drugs (e.g., colchicine and benomyl) induces mitotic arrest by activating a feedback control that senses improperly assembled spindles and prevents MPF inactivation. In budding yeast, mutations called *mad* (mitotic arrest deficient) and *bub* (budding uninhibited by benomyl) destroy this spindle-dependent feedback control of MPF inactivation by cyclin B degradation. When these mutants are treated with microtubule-depolymerizing drugs, anaphase occurs before the chromosomes are properly aligned, resulting in large-scale chromosome loss and death of the daughter cells. Curiously, these mutants are viable in the absence of drug. How can this be?

Answers

1.   APC is inactivated in G1 by G1Cdk-mediated phosphorylation. This complex is activated in mitosis; it degrades the anaphase inhibitor and mitotic cyclins.

For further study, see text section "Overview of the Cell Cycle and Its Control," p. 498.

2. Yeast cells carrying either temperature-sensitive (ts) or cold-sensitive (cs) *cdc* mutations first are transformed with a human cDNA library. Each transformed cell contains a single vector containing unique human genes. When the transformed cells are plated at the nonpermissive temperature, only those cells containing the wild-type, human homolog of the mutated gene will replicate. The plasmid containing the complementing gene can be isolated and the gene analyzed.

For further study, see text section "Overview of the Cell Cycle and Its Control," p. 499 and Figure 13-4.

3. The first few hours of *Xenopus* development show essentially synchronous divisions that have only two alternating phases, S and M. As a result, experiments can be completed in a (relatively) short period of time. As well, this abbreviated cycle occurs independently of the nucleus, so that the events regulating M phase *only* can be studied without interventions to eliminate other phases.

For further study, see text section "Biochemical Studies with Oocytes, Eggs and Early Embryos," p. 502.

4. The cyclic activity of MPF would be restored by adding the mRNA of *S. pombe* cdc13. The protein translated from this mRNA has homology to both *Xenopus* and sea urchin cyclin B. cdc13 complexed with *S. pombe* cdc2 forms this organism's MPF.

For further study, see text section "Biochemical Studies with Oocytes, Eggs and Early Embryos," pp. 502–503 and Figure 13-7. See also text section "Genetic Studies with *S. pombe*," p. 507.

5. The three-dimensional structure of the kinase alone shows that the top of a flexible region called the T loop blocks access of ATP to the binding site of the substrate. This observation explains why Cdk that is not bound to cyclin has no enzymatic activity. Binding of cyclin to the Cdk causes a change in the position of the T loop, exposing the Cdk active site. Phosphorylation of Thr-161 in the T loop causes a further conformational change that increases the affinity for substrate by one-hundred fold. The inhibitory Tyr-15 is located close to the ATP binding site. Phosphorylation of Tyr-15 prevents binding of ATP dues to electrostatic repulsion between the phosphates. This inhibition occurs even when Cdk is complexed with cyclin.

For further study, see text section "Genetic Studies with *S. pombe*," pp. 508–509 and Figure 13-14.

6. Dephosphorylation events during mitosis include protein phosphatases removing the regulatory phosphates from lamins A, B, and C, permitting reassembly of the nuclear laminae of the two daughter cell nuclei. When MPF activity falls during anaphase, a constitutive phosphatase dephosphorylates inhibitory sites on myosin light chain, allowing cytokinesis to proceed.

For further study, see text section "Molecular Mechanisms for Regulating Mitotic Events," pp. 514–517 and Figure 13-21.

7. Pre-replication complexes assemble at origins early in G_1 phase but are held in check until the inhibitory action of Sic1 is relieved by its degradation. Pre-initiation complexes contain Cdc28-Clb5 and Cdc28-Clb6 which, when active, initiate DNA replication and prevent assembly of additional pre-replication complexes from forming during S, G_2, and through metaphase of M.

For further study, see text section "Genetic Studies with *S. cerevisiae*," pp. 522-524 and Figures 13-25 and 13-27.

8. Within an organism, regulation of cell growth is essential to both determining the size and the shape of tissues during development and maintaining them during adulthood. Within organs, cell are differentiated to carry out a particular function. Restriction of cell division, as occurs in G_0, is a way to fulfill these requirements.

For further study, see text section "Cell Cycle Control in Mammalian Cells," p. 510.

9. Checkpoint controls, which prevent progression through the cell cycle when the previous step is incomplete or incorrect, have been demonstrated at several points. Two checkpoints respond to damaged DNA. These are at G_1 and G_2 and occur through the action of p53. DNA damage stabilizes p53, which acts as a transcription factor for a gene encoding p21[CIP]. p21[CIP] binds to and inhibits all cyclin-Cdk complexes, thereby arresting cells in G_1 and G_2. Another checkpoint control arrests cells containing replicated DNA with double-stranded breaks in G_2, thereby preventing unequal chromosome segregation to daughter cells. Unreplicated DNA prevents a rise in MPF and entry into mitosis, again by an unknown mechanism; this leads to arrest in G_2. Finally, if the spindle does not form properly, degradation of MPF is prevented and cells are arrested in mitosis. Yeast *bub* and *mad* mutants are defective in this checkpoint, and sequencing of the *BUB* gene indicates that it encodes a protein kinase.

For further study, see text section "Checkpoints in Cell-Cycle Regulation," pp. 529–532 and Figures 13-34 and 13-36.

10a. The DNA content of cells in G_0 and G_1 is identical (*2n* for diploid cells); thus the peak at 1 represents cells in both these phases. Likewise, the DNA content of cells in G_2 and M is identical (*4n* for diploid cells); thus the peak at 2 represents cells in both these stages. Cells in the S phase have an intermediate DNA content, corresponding to different amounts of replication. In this cell population, then, the fewest cells are in S; the most are in G_0+ G_1; and an intermediate number are in G_2 + M.

10b. See Figure 13-7.

10c. The FACS assay is incapable of distinguishing G_0 from G_1 or G_2 from M, because the DNA content of the respective cell cycle phase pairs is identical (see answer 10a).

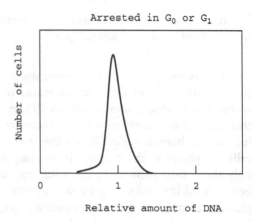

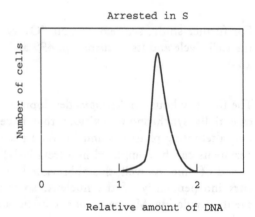

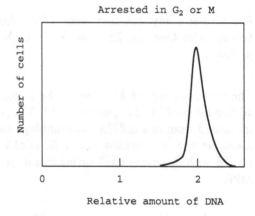

Figure 13-7

For further study, see text section "Overview of the Cell Cycle and Its Control," p. 496 and Figure 13-1.

11a. Peak 1 represents the cells' first DNA replication after the radioactive thymidine pulse; thus it corresponds to cells that were in the S phase during the pulse. Peak 2 represents the second DNA replication.

11b. These results must be from an unsynchronized mammalian cell culture. The chief tip-off is the shape and width of the first labeled mitotic cell peak. If a synchronized cell population were pulsed for 30 min with radioactive thymidine during S, this peak would increase abruptly in a step-like manner as the cells progressed through the remaining portion of S and then through G_2 and M. The peak also would decrease abruptly as the cells exited M. Since M is only 1 h in these cells, the width of the peak would be narrow if the cells were synchronized. The observed peak, in contrast, is broad, and is spread over ≈12 h, indicating that cell culture was unsynchronized. Note that a 30-min [^{3}H]thymidine pulse would result in no labeling of synchronized cell populations in G_1, G_2, or M.

For further study, see text section "Overview of the Cell Cycle and Its Control," p. 496 and Figure 13-1.

12a. These data suggest that the cytosolic extract is the source of licensing factor for the second round of replication. Treatment of the newly replicated sperm nuclei in experiment 1 (addition of MPF, which causes disassembly of the nuclear envelope) and experiment 3 (permeabilization) leads to a permeable nuclear envelope and hence free access of cytosolic molecules to the chromosomes; both of these treatments supported DNA re-replication. In contrast, the chromosomes in the newly replicated nuclei in experiment 2 are inaccessible to cytosolic molecules; in this case, no re-replication occurred.

12b. These experiments suggest that licensing factor is found in two different intracellular locations, the cytoplasm and the nucleus, as illustrated in Figure 13-8. Cytoplasmic licensing factor is stable, but lacks access to the nucleus until mitosis. At mitosis, the nuclear envelope disassembles and licensing factor from the cytoplasm binds to mitotic chromosomes. As the nuclear envelope re-forms, sufficient licensing factor is trapped within the nucleus to support DNA replication in the next S phase, but it is consumed in the process. The nuclear licensing factor present in the nuclei of the added sperm supports the first round of replication depicted in Figure 13-3; however, a second round of replication can occur only if mitosis occurs or the nuclear envelope is permeabilized.

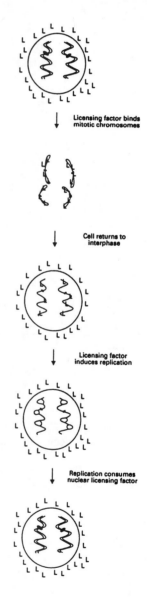

Licensing factor binds
mitotic chromosomes

Cell returns to
interphase

Licensing factor
induces replication

Replication consumes
nuclear licensing factor

Figure 13-8

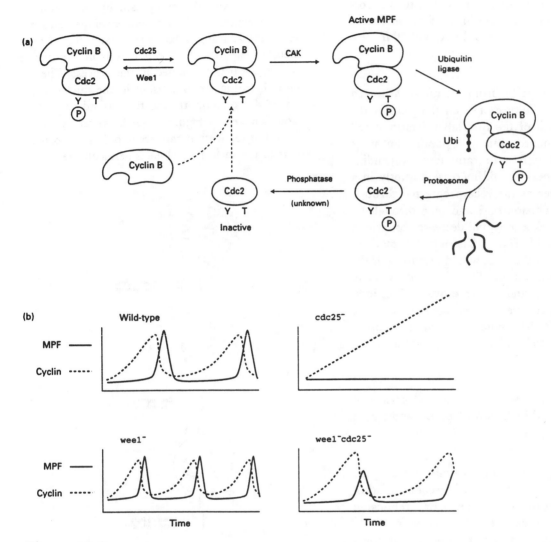

Figure 13-9

For further study, see text section "Biochemical Studies with Oocytes, Eggs and Early Embryos," p. 502.

13a. See Figure 13-9a. CAK = Cdc2-activating kinase.

13b. See Figure 13-9b. Wee1 is a tyrosine kinase that phosphorylates the Tyr-15 residue in the Cdc2 subunit of MPF, thereby inactivating it. Cdc25, a phosphatase, has the opposite effect, removing the inhibitory phosphate from Tyr-15. Subsequent phosphorylation of Thr-161 by the threonine kinase CAK generates catalytically active MPF. In the absence of Cdc25, the inactive phosphorylated form accumulates; as a result, no cycling of MPF activity or cyclin B level, and hence no cell divi-

sion, occurs. In the absence of Wee1, its inhibition of MPF is relieved, so that cycling and cell division are accelerated. Because of the opposite effects of Cdc25 and Wee1, absence of both proteins results in approximately normal (wild-type) cycling of MPF activity and cyclin B levels.

For further study, see text section "Genetic Studies with *S. pombe*," pp. 507–508 and Figure 13-12.

14a. The first cell cycle after shifting *wee1*[ts] mutants to the elevated nonpermissive temperature is shorter than normal. Since a smaller cell size is sufficient to meet the size threshold for mitosis, the duration of the G_2 phase is reduced, as illustrated in Figure 13-10.

14b. Subsequent cell cycles will be of normal duration. Because a *wee1* [ts] cell is smaller than normal as it exits mitosis, a longer G$_1$ phase is required for the cell to reach the threshold size required to pass START. The G$_2$ phase will remain shortened, since a small cell meets the threshold size for mitosis in these mutants (see Figure 13-10).

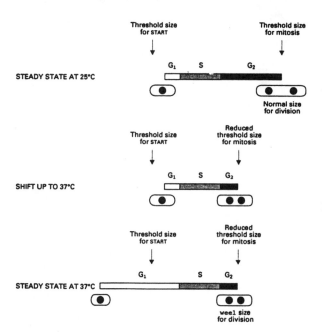

Figure 13-10

For further study, see text section "Genetic Studies with *S. pombe*," pp. 507–508 and Figure 13-12.

15. Nocodazole causes arrest in M by preventing spindle assembly. The cells first are incubated in the presence of nocodazole for a sufficient time period to permit progression through the proceeding phases of the cell cycle. The incidence of mitotically arrested cells may be monitored by the presence of condensed chromosomes and nuclear envelope breakdown, characteristic microscopic traits of a mitotic cell. After all, or nearly all, the cells are in M, the nocodazole is removed; functional spindles then are produced and the cells proceed to cycle in synchrony.

For further study, see text section "Molecular Mechanisms for Regulating Mitotic Events," pp. 512–513.

16. Clone C is the wild-type clone: It is not temperature sensitive in growth and cells within the population have an appearance typical of all phases of the cell cycle. Clone A is the *cdc* (cell division cycle) mutant. The cells appear uniform in size and shape, the result expected if the cells were arrested in a single phase of the cell cycle. Clone B is the non-*cdc* mutant: Although the cells are temperature sensitive in growth, they are heterogeneous in appearance at the elevated temperature (unlike *cdc* mutants).

For further study, see text section "Genetic Studies with *S. cerevisiae*," pp. 517-518 and Figure 13-22.

17a. The data show that the level of mRNA encoded by early-response genes begins to decline before mRNA encoded by delayed-response genes accumulates. This finding suggests that expression of the delayed-response genes most likely has nothing to do with shutdown of the early-response genes.

17b. The finding that cycloheximide prevents the shutdown of transcription of early-response genes indicates that translation of the early-response mRNA into protein is required for this shutdown. A reasonable hypothesis based on this finding is that one of the early-response proteins functions, either directly or indirectly, to shut down transcription of this set of genes. For example, an early-response protein might be a transcriptional repressor or it might inactivate a transcription factor required for expression of early-response genes.

For further study, see text section "Cell Cycle Control in Mammalian Cells," pp. 526-527 and Figure 13-30.

18. In both wild-type cells and the mutant cells, the events that lead to MPF inactivation normally take longer than spindle assembly, so spindle assembly generally precedes MPF inactivation. In the presence of microtubule-depolymerizing drugs, spindle assembly is delayed; in this situation, the spindle-dependent feedback control of MPF inactivation is crucial. The *mad* and *bub* mutants lack this control and thus do not arrest in mitosis in the presence of drug. However, in the absence of microtubule-depolymerizing drug, the mutant cells behave as normal cells, with the spindle forming before MPF inactivation, leading to normal mitosis.

For further study, see text section "Checkpoints in Cell-Cycle Regulation," pp. 530-531.

14 Gene Control in Development

PART A: *Chapter Summary*

The ultimate goal of developmental biologists is to unravel the mystery of how a fertilized egg is transformed into a complex multicellular organism. This process requires execution of a complex developmental program whereby specific genes are activated in a precise time-sequence and in the correct location, generating different types of tissues and the specific cell types composing them. Classical and molecular geneticists have discovered numerous genes that participate in the highly regulated programs that result in the development of plants and animals. Understanding the molecular basis for the action of such genes is one of the most actively studied areas in all of Biology.

By far the most prominent mechanism for regulating the production of different proteins in different cells entails cell type–specific DNA-binding proteins that either activate or repress gene transcription. This chapter takes a closer look at the spatial and temporal control of gene expression during development. Genetic and molecular studies show that different cells express different sets of genes based on their developmental history, their patterns of cell division, their position in the developing organism, and their interactions with other cells. Several well-studied cases of differential gene transcription which specify different cell types in yeast, animals, and plants are presented to illustrate the general transcription-control strategies that regulate development.

PART B: *Reviewing Concepts*

14.1 *Cell-Type Specification and Mating-Type Conversion in Yeast*
(lecture date _____)

1. Both protein-DNA interactions and protein-protein interactions are important in specifying the characteristics of the three yeast cell types: haploid a, haploid a, and diploid a/a cells. Which interactions are important in regulating cell type-specific genes in a haploid cells and α haploid cells?

2. How is the production of haploid-specific proteins, as well as a- and α-specific proteins, repressed in diploid cells.

3. How is mating-type conversion restricted to the mother cell after mitosis?

4. How does chromatin structure influence mating-type conversion?

14.2 *Cell-Type Specification in Animals*
(lecture date _____)

5. Exposure of C3H 10T1/2 cells to 5-azacytidine, a nucleotide analog, is a model system for muscle differentiation. How was 5-azacytidine treatment used to isolate the genes involved in muscle differentiation?

6. A critical event in differentiation is the transcription of cell type-specific genes. Both MyoD and the E2A protein, as well as other myogenic proteins, contain the helix-loop-helix (HLH) motif, which binds to a 6-bp DNA sequence called the E box. Although many copies of the E box probably are located throughout eukaryotic genomes associated with a variety of genes, MyoD and E2A only activate muscle-specific genes. Describe the probable mechanism of this specificity of MyoD and E2A action and the experiments that led to the elucidation of this mechanism.

7. In multicellular organisms, development of specific cell types expressing unique proteins occurs in a stepwise fashion and depends on various developmental genes. Define the two basic steps in cell-type specification and identify the genes involved in each step during myogenesis in mammals and neurogenesis in *Drosophila*.

14.3 *Anteriorposterior Specification during Embryogenesis* (lecture date _____)

8. A central concept of developmental biology is that the developmental fate of cells is determined by certain molecules called morphogens. Explain how morphogens are thought to determine different cell fates.

9. Which maternal genes function as anterior and posterior determinants during *Drosophila* embryogenesis and how do their encoded proteins act to establish the early anterior/posterior patterning of the embryo?

10. What are the molecular controls that result in a gradient of Hunchback (Hb) protein?

11. Give an example in which a *Drosophila* probe has been used to find a mammalian homolog of a *Drosophila* developmental gene and discuss the importance of these findings for developmental biology.

14.4 *Specification of Floral-Organ Identity in Arabidopsis* (lecture date _____)

12. Predict the phenotypes of *Arabidopsis* mutants defective in both A and B function and B and C function, respectively.

13. Two genes, *def* and *glo*, are required for B function in snapdragon flower identity. What is the phenotype of plants mutant in either *def* or *glo*, and what kind of proteins would def and glo encode? How do such proteins function in floral development?

PART C: *Analyzing Experiments*

14.1 *Cell-Type Specification and Mating-Type Conversion in Yeast*

14. (•) Mating-type conversion is restricted to mother cells after mitosis. The presence of Ash1 protein, an inhibitor of Swi5, is thought to control the activity of Swi5 in promoting HO transcription in mother cells. Researchers have studied the control of Ash1 expression to discern how the presence of Ash1 controls mating-type conversion. The data in Figure 14-1 was derived from experiments utilizing a fluorescently labeled antisense *Ash1* RNA probe which was hybridized to yeast cells undergoing mitosis. Bound probe can be visualized with fluorescent microscopy to detect sense *Ash1* mRNA in wild-type and mutant yeast cells containing a deletion of the *Ash1* gene (ash1Δ). The top panel shows fluorescent detection of *Ash1* mRNA, the middle panel utilizes a fluorescent nuclear stain to indicate the presence of nuclei within the same two yeast cells, and the bottom panel shows the light microscopic image of the same two yeast cells.

a. Is the RNA expression pattern of *Ash1* in wild-type yeast consistent with what is known about mating-type conversion?

b. What is the difference in *Ash1* expression in wild-type vs. ash1Δ mutants?

c. *Ash1* RNA accumulation was also investigated in various mutant strains of yeast (Figure 14-2). 3" UTRΔ is a mutant containing a deletion of the 3' UTR of the *Ash1* gene. Two other mutants, bniΔ and myo4Δ, contain mutations in genes involved in promoting actin-filament formation and myosin respectively. Explain why *Ash1* mRNA accumulates in both mitotic products in the 3'UTR Δ mutant.

d. What can be inferred about the role of the actin filaments and myosin in mating-type conversion?

e. Would a-specific or α-specific genes be transcribed in the progeny of the bniΔ mutant cells? Would they be expressed in ash1Δ mutant cells?

Wild type   ash1Δ

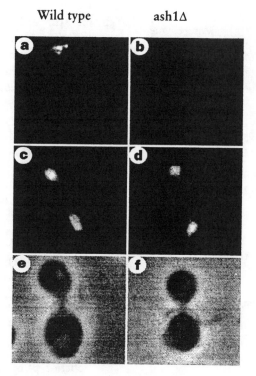

Figure 14-1

3'UTR_Δ   bniΔ   myo4Δ

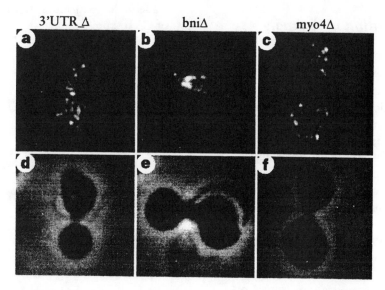

Figure 14-2

(a)

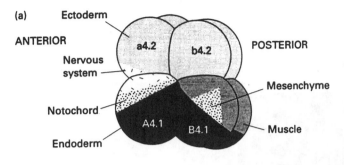

(b)

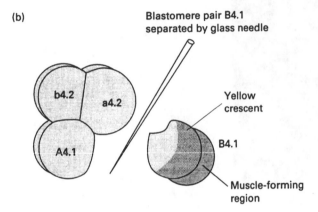

Figure 14-3

14.2 *Cell-Type Specification in Animals*

15. (••) In vertebrates, regionalization is determined by interactions between cells within the embryo; this is termed *conditional specification*. In *Drosophila* embryos, early regionalization is determined by interactions between cytoplasmic regions prior to cellularization of the embryo; this is termed *syncytial specification*. In *Drosophila* and most other insects, conditional specification only occurs after cellularization. A third pattern is found in most invertebrates. In these organisms, regionalization is determined by cellular acquisition of cytoplasmic molecules present in different portions of the egg; this is called *autonomous specification*. Figure 14-3a shows the 8-cell stage of the embryo of a tunicate, an invertebrate that exhibits autonomous specification. In this organism, the yellow plasm of B4.1 cells contains a component initiating muscle-specific development which is similar to the MyoD1/myogenin family of transcription factors.

14.3 *Anteriorposterior Specification during Embryogenesis*

16. (••) You have constructed two sets of gene fusions. In one, the *Antennapedia* (*Antp*) promoter is fused with CAT, which functions as a reporter gene; in a second, the actin promoter, a strong constitutive promoter, is fused with either the *fushi tarazu* (*ftz*) gene or the *Ultrabithorax* (*Ubx*) gene. The gene fusions are then expressed singularly or in combination in cultured *Drosophila* cells. The results of the specific fusions and their expression/co-expression are summarized in Figure 14-4.

a. Based on these results, are the proteins encoded by *ftz* and *Ubx* positive or negative regulators of *Antp* transcription?

b. In a 4.5 hour-old embryo (as in Figure 14-32 in the textbook), ftz is expressed in a seven stripe pattern.

   Reconcile this transcription pattern with the data in Figure 14-4.

c. Based on the results in Figure 14-4, predict the phenotype of an *Ubx* mutant fly.

d. If the constructs in Figure 14-4 were examined in a Drosophila cell culture made from an extradenticle (*exd*) mutant, what would the level of CAT expression be?

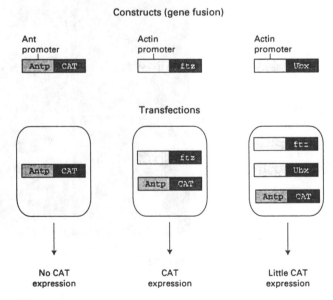

Figure 14-4

17. (•••) In temperature-shift experiments with *Drosophila* carrying a temperature-sensitive mutation in the *hedgehog* (*hh*) gene, you identify two periods during which *hedgehog* function is critical for development of a normal phenotype: the first occurs 2–10 h after fertilization and the second occurs at pupation. To learn more about regulation of the *hedgehog* gene and the function of its encoded protein, you determine the level of *hedgehog* mRNA by Northern blotting of extracts of wild-type *Drosophila* embryos and adults. Your results are shown in Figure 14-5.

   a. Do the data in Figure 13-8 suggest that *hedgehog* mRNA, like *bicoid* mRNA, has a highly polarized distribution in the egg?

   b. Is Hedgehog protein metabolically stable? Explain.

   c. Would mutations that lengthen or shorten the metabolic stability of Hedgehog protein affect *Drosophila* development?

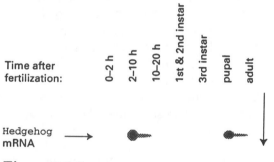

Figure 14-5

14.4 *Specification of Floral-Organ Identity in Arabidopsis*

18. (•••) *Ranunculus* is a flowering plant whose genus is distantly related to *Arabidopsis* and other angiosperms. Various species of *Ranunculus* have a flower containing whorls of sepals (Se) petals (Pet), stamens, (Sta) and carpels (Car). Within this arrangement, petals themselves can comprise many whorls and can be divided into different developmental stages from immature petals (Pet4) to mature petals

(Pet8). To determine if the ABC Model of flower development holds true for various species of *Ranunculus*, AP-3 and *Pistillata (PI)* genes from these species were cloned and expression of their cognizant RNAs were measured in Northern blot experiments. Figures 14-6 and 14-7 show the results of these experiments. Use this data to answer the following questions:

   a. What is the expected pattern of expression of *AP3/PI* genes if *Ranunculus* species utilize these genes as does Arabidopsis?

   b. Is the pattern of *AP-3/PI* gene expression in *Ranunculus bulbosus* (buttercup) indicative of conservation of function of *AP3/PI* genes (see Figure 14-6)? Why or Why not?

   c. Is the pattern of *AP-3/PI* gene expression in *Ranunculus ficaria* (lesser celandine) indicative of conservation of function of *AP3/PI* genes (see Figure 14-67)? Why or why not?

   b. Would genetic analysis of the *AP-3* and *PI* genes in *Ranunculus* yield the same phenotypes as loss-of-function B mutants in Arabidopsis? Why or why not?

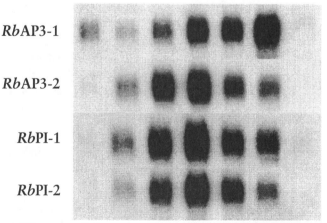

Figure 14-6

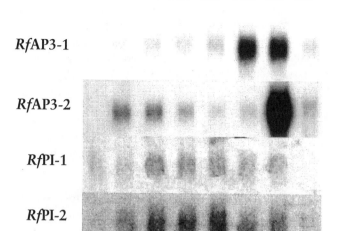

Se Pet4 Pet5 Pet6 Pet7 Pet8 Sta Car

*Rf*AP3-1

*Rf*AP3-2

*Rf*PI-1

*Rf*PI-2

Figure 14-7

Answers

1. In a cells, synthesis of mRNAs encoding a-specific proteins is stimulated by binding of a dimer of MCM1, a constitutive transcription factor, to the P site in the upstream regulatory sequences (URSs) of a-specific genes. In the absence of α1, no activation of α-specific genes occurs. In α cells, transcription of α-specific genes is activated by the simultaneous binding of two proteins to the α-specific URSs, which comprise two sites P and Q. Activation requires binding of dimeric MCM1 to P and binding of α1 protein to Q. Transcription of a-specific genes in α cells is repressed by the cooperative binding of dimeric MCM1 to the P site and dimeric α2 to the flanking α2 sites in a-specific URSs.

   For further study, see text section "Combination of DNA-Binding Proteins Regulate Cell-Type Specification in Yeast," p. 538, and Figure 14-1.

2. In diploid cells, transcription of haploid cell-specific genes is repressed by the α2-a1 heterodimer. This heterodimer also represses synthesis of α1 mRNA; in the absence of α1, a-specific genes are not transcribed. Transcription of a-specific genes in diploid cells is repressed by the MCM1-α2 complex by the same mechanism as in α cells.

   For further study, see text section "Combination of DNA-Binding Proteins Regulate Cell-Type Specification in Yeast," p. 538, and Figure 14-1.

3. In mother cells, Swi5 protein synthesized during the previous cell cycle binds to two sequences within URS1 and promotes transcription of the HO locus. The HO encoded gene product cleaves a specific site at a silent MAT locus which allows for transfer of a HML or HMR gene into the central actively transcribed MAT locus. In daughter cell, Swi 5 is inactive, presumably due to the presence of Ash1 in these cells; therefore, HO is not transcribed and mating-type conversion does not take place.

   For further study, see text section "Multiple Regulation of HO Transcription Controls Mating-Type Converison," p. 541, and Figure 14-6.

4. The HO locus can be maintained in a transcriptionally inactive form by chromatin. The SIN genes are involved in maintenance of this repressive state, which does not allow the postively acting Swi 1, 2 and 3 proteins to bind. As well, silencer elements are involved in repressing the silent copies of HMLR and HMR at the MAT locus. These silencer elements assemble into higher-order chromatin structures which are inaccessible to the transcriptional machinery.

   For further study, see text section "Silencer Elements Repress Expression at HML and HMR," p. 542.

5. Exposure of C3H 10T1/2 cells to 5-azacytidine is thought to induce muscle differentiation by incorporation of this compound, which cannot be methylated, into DNA. As a result, genes that previously had been inactivated by methylation are re-activated, and a different phenotype is expressed. The first step in isolating the genes involved in muscle differentiation was to demonstrate that DNA isolated from cells treated with 5-azacytidine, called azamyoblasts, could transform untreated C3H 10T1/2 cells into muscle. The mRNAs isolated from azamyoblasts were then converted to cDNAs and subjected to subtractive hybridization with mRNAs extracted

from untreated cells. The azamyoblast-specific cDNAs then were used as probes to screen an azamyoblast cDNA library. The genes isolated by this procedure were tested for their ability to promote muscle differentiation by transfecting them into C3H 10T1/2 cells and assaying for the muscle protein myosin with an immunofluorescence assay.

For further study, see text section "Myogenic Genes Were First Identified in Studies with Cultured Fibroblasts," p. 544, and Figure 14-10.

6. The E box is present in multiple copies in the enhancers of muscle-specific genes. MyoD activates transcription of these genes when it binds cooperatively to two or more of these sites as a heterodimer with another protein, E2A. The importance of the E2A protein was shown by inhibiting E2A production in C3H T101/2 cells with antisense RNA; cells treated in this way did not undergo myogenesis in the presence of 5-azacytidine. Wild-type E2A cannot induce conversion of CH3 T101/2 cells into myotubes. However, a variant E2A in which the DNA-binding sites are mutated to mimic those found in MyoD does induce myogenesis in C3H T101/2 cells. This result suggests that when MyoD or the variant E2A is bound to E boxes associated with muscle-specific genes, the protein assumes a conformation that permits transcriptional activation. In contrast, when E2A or MyoD bind to E boxes associated with genes encoding nonmuscle proteins, their conformation may not be compatible with transcriptional activation of these genes.

For further study, see text section "Myogenic Proteins Are Transcription Factors Containing a Common bHLH Domain" and "MEFS Function in Concert with MRFs to Confer Myogenic Specificity," p. 546.

7. The two basic steps in myogenesis, neurogenesis, and other developmental pathways are determination and differentiation. During determination, precursor cells become committed to a particular developmental pathway but do not yet exhibit the characteristics of differentiated cells. During differentiation, committed cells undergo further changes in gene expression resulting in production of cell type-specific proteins. Determination in mammal-

ian myogenesis depends on expression of the *myoD* and *myf5* genes, and determination in *Drosophila* neurogenesis depends on expression of the *achaete* and *scute* genes. The proteins encoded by these genes function in conjunction with general transcription factors (E2A for mammals and Da for *Drosophila*) to activate the mammalian *myogenin* and *Drosophila asense* genes, which are required for differentiation of muscle cells and sensory hairs, respectively.

For further study, see text sections "Embryonic Somites Give Rise to Myoblasts, the Precursors of Skeletal Muscle Cells," p. 535; "A Network of Cross-Regulatory Interactions Maintains the Myogenic Program," p. 549; and "Neurogenesis Requires Regulatory Proeins Analogous to bHLH Myogenic Proteins," p. 550, and Figure 14-17.

8. A morphogen can elicit different cellular responses depending on its concentration. Within the embryo, morphogens are distributed along concentration gradients, being present at higher concentrations in some regions than in others. Each morphogen exhibits a finite set of threshold concentrations associated with various cellular responses to it: above the threshold concentration, one response occurs; below the threshold, another response. Morphogens affect the developmental fate of cells by activating or repressing transcription of specific genes or by affecting translation of specific mRNAs. The sensitivity of cells to a particular morphogen may be mediated by the affinity of the morphogen for its binding sites within the genome. This has been demonstrated experimentally in the case of Bicoid protein regulating transcription of the zygotic *hunchback* gene.

For further study, see text section "Morphogens Regulate Development as a Function of Their Concentration," p. 556.

9. The anterior determinant is the *bicoid* gene. The *bicoid* mRNA is produced by the mother, deposited in the oocyte, and becomes localized in the anterior portion of the embryo. After Bicoid protein is synthesized, it diffuses posteriorly, establishing an anterior → posterior Bicoid concentration gradient. Bicoid protein stimulates transcription of the zygotic *hunchback* gene in the early embryo, generating a gradient of zygotic *hunchback* mRNA. The posterior determinant is the

nanos gene. Since the maternal *nanos* mRNA is localized in the posterior region of the embryo, a posterior → anterior gradient of maternal Nanos protein is established following translation of the mRNA. Since Nanos inhibits translation of *hunchback* mRNA, the uniformly distributed, maternally derived, *hunchback* mRNA is not translated in the posterior region of the embryo. Thus, as a result of the opposite gradients of Bicoid and Nanos protein, synthesized from maternal mRNAs, an anterior → posterior gradient of Hunchback protein is established.

For further study, see text section "Maternal *bicoid* Gene Specifies Anterior Region in *Drosophila*," p. 557; and "Maternally Derived Inhibitors of Translation Contribute to Early *Drosophila* Patterning," p. 558.

10. Maternal *Hb* RNA is uniformly distributed throughout the embryo, but these RNAs do not get translated in the posterior region due to repression by *Nanos*. Zygotic *Hb* RNA is under control of the Bicoid protein, which is localized to the anterior region. Both of these controls together ensure that Hb protein is only present at the anterior end of the embryo.

For further study, see text section "Maternally Derived Inhibitors of Translation Contribute to Early *Drosophila* Patterning," p. 558.

11. The mammalian homologs, called the Hox complex (Hox-C), of the *Drosophila* homeotic genes contained in *ANT-C* and *BX-C* were found by screening with a synthetic oligonucleotide probe encoding the DNA-binding homeobox domain, which is found in all the structural genes in the *Drosophila* HOM-C complex. The conservation of these DNA-binding motifs suggests that development in insects and mammals follows similar patterns and that discoveries using one model system may be transferable to another, thereby simplifying the task of developmental biologists, who hope to define completely the molecular events causing differentiation.

For further study, see text section "Mammalian Homologs of *Drosophila* ANT-C and BX-C Genes in Four Hox Complexes," p. 568.

12. An *Arabidopsis* double mutant defective in both A and B function will contain flowers with whorls of carpels, carpels, carpels, and carpels because all information required to make sepals, petals and stamens is lost with loss of A and C function. An *Arabidopsis* double mutant defective in both B and C function will contain flowers with whorls of sepals, sepals, sepals, and sepals, because information required for petals, stamens and carpels is lost with loss of B and C function.

For further study, see text section "Three Classes of Genes Control Floral-Organ Identity," p. 572.

13. Snapdragon flowers mutant in either *def* or *glo* contain four whorls of sepals, sepals, carpels, and carpels only. A good guess is that Def and Glo proteins are MADS family transcription factors that specify petal and stamen identity in conjunction with an A and C gene respectively.

For further study, see text section "Three Classes of Genes Control Floral-Organ Identity" p. 572 and "Many Floral Organ-Identity Genes Encode MADS Family Transcription Factors," p. 573.

14a. Yes. Only one progeny cell of the mitotic event (i.e. the daughter cell) is known to express Ash1 protein. The slightly smaller daughter cell in this experiment can be seen to express *Ash1* mRNA whereas the mother cell does not. This result indicates that the differential regulation of Ash1 protein occurs at the level of transcription.

14b. In wild type yeast, *Ash1* mRNA becomes segregated to one mitotic product. In Ash1Δ mutants, there is no *Ash1* mRNA produced, so neither mitotic product contains *Ash1* mRNA.

14c. The simplest explanation is that the 3' UTR of *Ash1* mRNA is required for targeted destruction of *Ash1* mRNA in mother cells.

14d. Since mutation in genes required for actin filament formation and myosin production disrupt the normal segregation of *Ash1* mRNA, actin filaments and myosin must be required for the normal segregation of *Ash1*. It is possible that *Ash1* mRNA attaches to the actin cytoskeleton and myosin during the segregation process.

14e. Since one does not know the mating phenotype of the parent bniΔ mutant cell before mitosis, it is impossible to predict the phenotype of the progeny afterwards. However, if the parent bniΔ cell originally expressed a-genes, then both products would continue to express a-genes as no mating-type conversion could occur with *Ash1* mRNA present. In *Ash1* mutants, the opposite would occur: mating-type conversion would occur in both progeny cells resulting in a switch from a-specific to α–specific gene expression.

15a. Since the separated B4.1 cells from the muscle-forming region contain the morphogens needed for differentiation, they will develop independently of the rest of the embryo into muscle cells; the rest of the embryo will form a multicellular mass devoid of muscle cells. Experimentally this is the observed result.

15b. The effects on invertebrate development of transferring cytoplasm from B4.1 cells to a4.2 cells is difficult to predict. This transfer would introduce new morphogens into the a4.2 cells. Whether or not the concentration of transferred morphogen would be sufficient to activate expression of genes not normally expressed in a4.2 cells cannot be readily predicted. Likewise, the effect of this transfer on the normal gene expression of the a4.2 cells is difficult to predict. The transfer experiment would have to be performed and then analyzed to answer this question.

15c. Injected MyoD or Myf5 proteins would likely produce the same result as the transfer of cytoplasm from muscle progenitors (B4.1 cells). These proteins could stimulate muscle cell development in the recipient cells.

15d. If the MyoD activity in tunicates is conserved with that of mammals, it is possible that microinjected Id protein would dimerize with native MyoD forming inactive Id/MyoD heterodimers. The onset of muscle determination and differentiation could be blocked in this experiment.

16a. The Ftz protein is a positive regulator of the *Antp* promoter, whereas the Ubx protein is a negative regulator of the same promoter. This is indicated by the altered levels of reporter-gene expression under control of the *Antp* promoter when different constructs are placed in the same cell.

16b. If Ftz was the only regulator of Antp transcription, Antp would be co-expressed in a seven stripe pattern along with its transcriptional activator, Ftz. However, many genes are involved in the regulation of *Antp*, so the actual pattern is dependent on other genes including repressors such as *Ubx*. This eventually results in a single broad stripe of Antp expression corresponding to the second thoracic segment of the fly (see Figure 14-32b).

16c. If a repressor of Antp transcription such as Ubx is missing, then the first abdominal segment (where Ubx normally functions to pattern the embryo) will express the Antp protein. This will result in a homeotic conversion of the first abdominal segment to the third thoracic segment. Since the third thoracic segment specifies the wings, such mutants will have a second pair of wings.

16d. In an *exd* mutant background, *Hox* genes are expressed but do not carry out their normal functions. Therefore, the actin-ftz construct would still activate transcription of the Antp-CAT construct, but the actin-Ubx construct would not function as a repressor of Antp-CAT. As well, since recent studies have shown that *exd* may convert Hox proteins from repressors to activators, expression of Antp-CAT might become even greater in the presence of actin-Ubx than with actin-ftz alone.

17a. The absence of a hedgehog hybridization band in the lane corresponding to early embryos (0–2 h postfertilization) indicates that hedgehog mRNA is not present in the egg.

17b. The critical periods for hedgehog function identified in temperature-shift experiments with hedgehog mutants coincide with the peaks in hedgehog mRNA production in wild-type flies. This coincidence could only occur if hedgehog protein turned over fairly rapidly, with a half-life of 1 h or less. Thus the protein is metabolically unstable.

17c. Mutations that stabilize hedgehog would increase the amount of protein present and lengthen the duration of its presence within the embryo. Corresponding mutations that destabilized the protein would have the converse effect. In nature, hedgehog, like other developmental proteins, is needed at the right time and likely in the right amount for normal development to occur. Hence, mutations affecting hedgehog stability are likely to affect *Drosophila* development. Hedgehog is involved in establishing segment polarity in the early embryo (2–10 h postfertilization) and is critical for patterning the adult appendages during the pupal stage.

18a. According to the ABC model, the individual *AP-3* and *PI* genes should be expressed in petals and stamens only. Given that there are multiple whorls of petals in *Ranunculus*, *AP-3/PI* expression might differ in these different petal whorls and still conform to the ABC model.

18b. No. The pattern of expression of all genes except *RbAP3-1* is indicative of conservation of B function in this species. Even though levels of the *RbAP3-2*, *RbPI-1* and *RbPI-2* genes are differentially expressed in the various petal whorls they follow a developmental pattern of low expression in immature petals (Pet4) with rising expression in more mature petals (Pet6) and continuing expression in the stamen whorl. However, the *RbAP3-1* gene is expressed at a higher level in sepals than in immature petals, and this does not conform to the ABC model.

18c. No. Once again, one gene, *RfPI-1* is expressed in the sepals which is not consistent with the ABC model. As well, the *RfAP3-2* gene is expressed in immature petals and then becomes downregulated in more mature petals, before finally being expressed at high levels in the stamen whorl. Subsequently, this same gene is expressed at a lower degree in the carpels which is not consistent with the ABC model.

18d. It appears that both *Rb* and *Rf AP-3/PI* genes are functionally redundant. That is to say that multiple *AP3/PI* genes are expressed in the same tissues at the same time during flower development. This redundancy makes it unlikely that a mutation in any single *AP3* or *PI* gene would result in a total loss of B function, as occurs in *Arabidopsis*, a plant with a single *AP-3* and *PI* gene. Without a total loss of B function, it would be impossible to see a mutant phenotype.

15 Transport Across Cell Membranes

PART A: *Chapter Summary*

The plasma membrane is the essential selective permeability barrier between the cell and its extracellular environment. The selective permeability of the plasma membrane allows the cell to maintain a constant internal environment. Metabolites such as glucose, amino acids, and lipids readily enter the cell, metabolic intermediates remain in the cell, and waste components leave the cells. The phospholipid bilayer is essentially impermeable to most water-soluble molecules. Transport of glucose, amino acids, and ions into or out of cells requires different mixtures of transport proteins. Similarly, organelles within the cell contain in their membranes specific transport proteins that are essential to creating the internal environment of the organelle.

In animals, sheets of cells, termed epithelial cells, line all the body cavities. Epithelial cells frequently transport ions or small molecules from one side to the other. Those lining the small intestine transport products of digestion into the blood and those lining the stomach secrete hydrochloric acid into the stomach lumen. For epithelial cells to carry out these transport functions, their plasma membranes are organized into two discrete regions, each with specialized transport proteins. In addition, epithelial cells are interconnected by junctional complexes that lend rigidity to the cell sheet and prevent material from moving between cells and from one side to the other.

We consider first the protein-independent movement of small hydrophobic molecules across phospholipid bilayers and an overview of the various types of transport proteins present in cell membranes. Second, after characterization of each of the main types of transport proteins, we review how specific combinations of transport proteins in different subcellular membranes enable cells to carry out important physiological processes, including the maintenance of cytosolic pH, the transport of glucose across the absorptive intestinal epithelium, and the directed flow of water in both plants and animals. The same type of transport protein may be involved in quite different physiological processes. Finally we consider a range of specific experimental examples.

PART B: *Reviewing Concepts*

15.1 *Diffusion of Small Molecules across Phospholipid Bilayers*
(lecture date _____)

1. Which aspects of the selective permeability properties of a biomembrane may be attributed to its phospholipid components, and which to its protein components?

2. What effect does increasing the number of $-CH_2$- groups in a molecule have on the parameters included in Fick's Law?

15.2 *Overview of Membrane Transport Proteins* (lecture date _____)

3. Both uniporter proteins and ion channel proteins facilitate the movement of a single substance down a concentration gradient. Ion channels typically move 10^7–10^8 ions/s see while uniporters more typically move 10^2–10^4 molecules. Why are uniporters so much slower than ion channels?

4. Why is "cotransporter" a better term to describe a symport or antiport than "active transporter"?

5. How do artificial liposome and cell expression systems contribute to analyze the functional properties of membrane-transport proteins?

15.3 *Uniport-Catalyzed Transport*
(lecture date _____)

6. Glucose levels in the human bloodstream vary over the range of 3 to 7 mM. GLUT1 is a glucose uniport found in many cell types. It has a K_m of 1.5 mM and a V_{max} of 500 µmol glucose/ml packed cells/h. How does the rate of glucose transport by GLUT1 change, going from a glucose concentration of 3 mM, a starvation condition, to 5 mM, the normal condition, to 7 mM, a condition found after a feast?

7. What effect does a uniporter have on the chemistry of the transported substance?

15.4 *Intracellular Ion Environment and Membrane Electric Potential*
(lecture date _____)

8. In mammalian cells, the intracellular K^+ concentration is about 140 mM and the extracellular K^+ concentration is about 5 mM. Assuming that the plasma membrane is permeable only to K^+, calculate the potassium equilibrium potential E_K across the membrane.

9. How is the inside negative plasma membrane potential achieved in plant and fungal cells?

15.5 *Active Transport by ATP-Powered Pumps* (lecture date _____)

10. Why is the plasma membrane Na^+/K^+ ATPase considered to be an electrogenic pump?

11. Compare and contrast the substrate specificity of ABC pumps with that of P, F, and V class pumps.

12. Chemotherapy with one drug, (e.g. colchicine), frequently selects for cells resistant to several chemotherapeutic agents (e.g., colchicine, vinblastine, and adriamycin). What is the molecular basis of this phenomenon?

15.6 *Cotransport by Symporters and Antiporters* (lecture date _____)

13. Explain the relationship between HCO_3^- and antiporters in regulating cytosolic pH in animal cells?

15.7 *Transport across Epithelia*
(lecture date _____)

14. How is glucose transported from the lumen of the intestine across the intestinal epithelium into the blood?

15. Cholera toxin, produced by *Vibrio cholerae*, a pathogenic intestinal bacterium, causes an indirect reduction in the activity of the Na^+/K^+ ATPase in intestinal epithelial cells. This results in reduced uptake of small sugars and amino acids from the intestine. How is this reduction in uptake coupled to impaired Na^+/K^+ ATPase function?

15.8 *Osmosis, Water Channels, and the Regulation of Cell Volume*
(lecture date _____)

16. Frog oocytes and eggs, which have an internal salt concentration of 150 mM, do not swell when placed in water with a very low solute concentration. Erythrocytes which have a very similar internal solute concentration rapidly swell and burst in solutions of very low osmolarity. What could account for this difference?

17. The concentrations of solutes in the cytosol and vacuole of plant cells are generally much higher than those typical in the cytosol and lysosome of mammalian cells. Why do plant cell plasma membranes not undergo osmotic lysis?

PART C: *Analyzing Experiments*

15.2 *Overview of Membrane Transport Proteins*, 15.3 *Uniport-Catalyzed Transport*

18. (••) You have conducted a series of experiments measuring the rate of glucose uptake by erythrocytes from two different patients (DZ and KD) and a normal control group. The data are shown in Figure 15-1 in which the uptake rate (in µmol glucose/ml packed cells/h) is plotted as a function of the external glucose concentration.

a. Estimate the V_{max} and K_m for glucose uptake by erythrocytes from the control population and each of the patients.

b. What is the nature of the defect(s) in patient DZ?

c. Propose two explanations for the defect in patient KD.

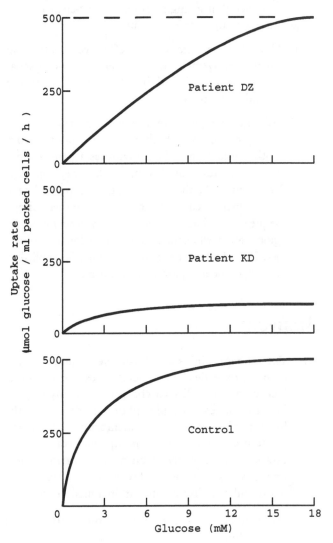

Figure 15-1

15.5 Active Transport by ATP-Powered Pumps

19. (•) When present in the culture medium, ouabain inhibits the Na⁺/K⁺ ATPase in cultured cells, but when the drug is microinjected into cells, it exerts no inhibitory effect. Although ouabain itself possesses no negative charge, ouabain treatment causes the α subunits of the Na⁺/K⁺ ATPase to become more negatively charged but does not alter the charge properties of the β subunits.

   a. On which side of the plasma membrane are the ouabain- and ATP-binding sites of the enzyme located?

   b. During the transport process, the Na⁺/K⁺ ATPase becomes transiently phosphorylated. What do these observations suggest about a likely mechanism for ouabain inhibition of this ion pump?

   c. Which of the two subunits, α or β, is most likely to be phosphorylated?

15.6 Cotransport by Symporters and Antiporters, 15.2 Overview of Membrane Transport Proteins, 15.7 Transport across Epithelia

20. (•••) The intestinal two-Na⁺/one-glucose symporter was sequenced by an expression cloning approach. Messenger RNA was expressed by microinjection into a frog oocyte. Three days following injection the level of Na⁺–dependent glucose transport was assessed. mRNA coding for the transporter was found to have a size of about 2.2 kilobases. A cDNA library corresponding to mRNAs of this appropriate size was created. Progressively smaller pools of the 2000 cDNA clones positive for Na⁺-dependent glucose transport were assayed. In the end a single clone was isolated and the cDNA sequenced. The deduced protein sequence was then analyzed to predict the membrane topology of the protein. The initial approach was based on the outcome of hydropathy plots. In hydropathy plots, the

hydrophobicity/hydrophilicity of amino acids are plotted relative to their position within a protein sequence. The hydropathy plot for the two-Na⁺/one-glucose symporter is presented in Figure 15-2. A current day understanding of the membrane topology of the two-Na⁺/one-glucose symporter is shown in MCB, page 598, Figure 15-18.

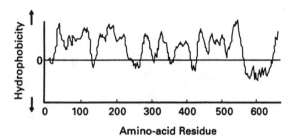

Figure 15-2

a. Expression cloning is one approach to screening cDNA libraries. Other approaches include antibodies or synthetic oligonucleotide probes derived from partial amino acid sequences. Based on the hydrophobicity plot and the fact that the intestinal two-Na⁺/one-glucose symporter accounts for less than 0.2% of plasma membrane protein, why might one predict that antibody or partial amino acid sequence derived oligonucleotide probe approaches would prove to be impractical? Note: The plasma membrane is typically about 0.5% of total cellular protein and there is about 2×10^{-4} µg of protein per cell.

b. From analysis of the hydrophobicity plot, how many transmembrane domains do you predict the intestinal two-Na⁺/one-glucose symporter to have?

c. From your prediction of transmembrane domains, do you expect the amino and carboxyl terminus of the protein to be on the same or different sides of the membrane?

d. In the actual publication in *Nature*, amino acids 314 through 334 were predicted to be a transmembrane domain. If modeled as an α helix, this portion of amino acids would be an amphipathic helix with a lysine residue (K) located approximately half-way along the helix. What thermodynamic problem is posed by K in this position? What is the likely solution to this problem?

15.7 *Transport across Epithelia*

21. (•••) The introduction of claudin as an expressed protein in mouse L cells is sufficient to cause the formation of morphologically identifiable tight junctions. The tight junctions form over extensive but relatively limited regions of cell-cell contact. Mouse L cells do not normally express tight junctions.

a. To what extent do these experiments provide evidence that a single protein is sufficient to produce a recognizable tight junction?

b. Should these junctions be effective in blocking nutrient movement across a cell layer or in separation of apical and basolateral cell surfaces? Explain why or why not.

c. An epitope tagging approach can be taken to ask the question of whether or not claudin from both contacting cells contributes to the formation of the junctional complex. In an epitope tagging approach, the epitope tagged expressed protein is engineered to include one or another epitope sequences coding for the antibody recognition regions for well-characterized monoclonal antibodies. Describe how the approach could be applied to the experimental question raised here?

Answers

1. The phospholipids establish the general permeability properties of biomembranes, for example, lack of permeability to charged versus hydrophobic substances. The phospholipids determine the hydrophobic core of the membrane. The membrane proteins as various pumps and transporters establish the selective transport of various ions, amino acids and sugars. Proteins can create local hydrophilic environments within a membrane.

    For further study, see text section "Diffusion of Small Molecules across Phospholipid Bilayers," p. 579.

2. An increasing number of -CH₂- groups increases the hydrophobicity of a molecule, the greater the hydrophobicity, the greater the value of K the partition coefficient. An increase in the value of K results in an increase in the value of P the permeability coefficient in Fick's Law.

For further study, see text section Diffusion of Small Molecules across Phospholipid Bilayers," p. 579.

3. Uniporter are slower than channels because they mediate a more complicated process. The transported substrate both binds to the uniporter and elicits a conformational change in the transporter. A uniporter transports one substrate molecule at a time. In contrast, channel proteins form a protein-lined passageway through which multiple water molecules or ions move simultaneously, single file, at a rapid rate.

For further study, see text section "Overview of Membrane Transport Proteins," p. 580.

4. Symporters and antiporters, like ATP pumps, mediate coupled reactions in which an energetically unfavorable reaction is coupled to an energetically favorable reaction. However, unlike pumps, neither symporters nor antiporters hydrolyze ATP or any other molecule during transport. Hence, these transporters are better referred to as cotransporters rather than active transporters. The term active transporter should be restricted to the ATP pumps where ATP is hydrolyzed in the transport process.

For further study, see text section "Overview of Membrane Transport Proteins," p. 580.

5. Both articial liposome and cell expression systems are experimental situations designed to test for the effect of a single gene product, i.e., protein. A specific transport protein may be extracted and purified. It may then be reincorporated into pure phospholipid bilayers, i.e., liposomes, and transport properties assessed. Alternatively, the protein may be expressed in a cell not otherwise expressing the transport protein. The differences in transport properties will be due to the newly expressed protein. In either case, the functional properties of the various proteins can be examined with little, if any, ambiguity.

For further study, see text section "Overview of Membrane Transport Proteins," p. 580.

6. This problem can be solved using the Michaelis equation:

$$v = V_{max} \frac{[C]}{[C] + K_m} \quad \text{(see MCB, p. 583)}$$

where [C] is the concentration of the transported molecule, in this case, glucose. By substituting the values for V_{max}, K_m, and the three glucose concentrations [C], the initial velocities in mmol glucose/ml packed cells/h can be calculated: at 3 mM glucose, $v = 333$; at 5 mM glucose, $v = 385$; and at 7 mM glucose, $v = 412$.

Hence, although the glucose concentration increases over a range of more than 2-fold v increases by only 24%.

For further study, see text section "Kinetics of GLUT1-Catalyzed Movement of Glucose," p. 583.

7. A uniporter, like all other transporters, catalyzes the rate of movement of ions or molecules across a membrane. It does not catalyze the making or breaking of covalent bonds, and hence the transported substance is unchanged in chemistry.

For further study, see text section "Uniport-Catalyzed Tranport," p. 582.

8. $E_K = -0.089$ V or -89 mV. The electric potential across a membrane due to differences only in the K^+ concentration can be calculated from the Nernst equation

$$E_K = \frac{RT}{ZF} \ln \frac{[K^+_{out}]}{[K^+_{in}]} \quad \text{(see MCB, p. 587)}$$

by substituting the following quantities: $R = 1.987$ cal/degree = mol; $T = 310$ K; $Z = 1$; $F = 23,062$ cal/mol = V; $[K^+_{out}] = 5$ mM; and $[K^+_{in}] = 140$ mM. Note that the electric potential across the membrane is described in reference to the exterior potential that is arbitrarily defined as zero. Thus the extracellular K^+ concentration is placed in the numerator.

For further study, see text section "The Membrane Potential in Animal Cells Depends Largely on Resting K⁺ Channels," p. 586.

9. In contrast to animal cells where resting K⁺ channels play the dominant role in generating the electric potential across the plasma membrane, plant cells generate the inside-negative membrane potential by transport of H⁺ ions out of the cell by an ATP powered proton pump.

For further study, see text section "The Membrane Potential in Animal Cells Depends Largely on Resting K⁺ Channels," p. 586.

10. The Na^+/K^+ ATPase is electrogenic because it pumps 3 Na^+ out of the cells for every 2 K^+ pumped into the cell. Hence, the Na^+/K^+ ATPase contributes directly to the charge inbalance across the membrane and the overall inside-negative membrane potential.

For further study see text section "The Na^+/K^+ ATPase Maintains the Intracellular Na^+ and K^+ Concentrations in Animal Cells," p. 593.

11. Members of the ABC (ATP binding cassette) class pumps may either transport ions or small molecules such as amino acids, sugars, vitamins, peptides, or hydrophobic molecules. F and V class pumps transport only protons (H^+) and P class pumps transport only one of four cations (H^+, Na^+, K^+, or Ca^{2+}).

For further study, see text section "Active Transport by ATP-Powered Pumps," p. 588.

12. The ABC class pump described in eukaryotes was the multidrug-resistance (MDR) transport protein known as *MDR1*. *MDR1* exports a large variety of relatively hydrophobic drugs from the cytosol to the extracellular medium. Hence cells that survive one drug because of the presence of *MDR1* are apt to be resistant to several drugs. The initial drug then serves as a selective agent for cells having *MDR1* in their plasma membranes.

For further study, see text section "Mammalian MDR Transport Protein," p. 595.

13. Three different cotransporters work together in a coordinated manner in most animal cells to regulate intracellular pH. A Na^+/H^+ antiport removes excess protons from the cell and hence raises pH. A $Na^+HCO_3^-/Cl^-$ cotransporter serves to raise pH. At the pH at which the cotransporter is active the imported bicarbonate anion combine with protons to produce CO_2 that diffuses out of the cell. A Cl^--HCO_3^- antiporter is active at slightly elevated pH and serves to lower pH by in effect exporting OH^- as HCO_3^- (= $OH^- + CO_2$).

For further study, see text section "Several Cotransporters Regulate Cytosolic pH," p. 600.

14. As glucose is moved across the intestinal epithelium, it is transported up a concentration gradient to enter epithelial cells and then down a concentration gradient to the blood. Transport of glucose into epithelial cells from the intestinal lumen is mediated by a two-Na^+/glucose symport protein. The energetically favorable cotransport of Na^+ powers glucose import. The GLUT 2 uniport protein facilitates the movement into the blood stream of glucose at the basolateral surface of epithelial cells. The Na^+/K^+ ATPase establishes the Na^+ across the epithelial cell membrane.

For further study, see text section "Transepithelial Movement of Glucose and Amino Acids Requires Multiple Transport Proteins," p. 602.

15. Sugar and amino acid uptake from the intestine is mediated by symport proteins located in the apical membrane of intestinal epithelial cells. These symporters use the cotransport of Na^+ down its electrochemical gradient to drive sugar and amino acid movement up a concentration gradient. Impaired functioning of the Na^+/K^+ ATPase in the basolateral membrane results in a decreased Na^+ gradient because this pumps Na^+ out of the epithelial cells into the blood. As a result of the decreased Na^+ gradient, the ability of the epithelial cells to take up sugars and amino acids against a concentration gradient is reduced.

For further study, see text section "Transepithelial Movement of Glucose and Amino Acids Requires Multiple Transport Proteins," p. 602.

16. The permeability of the erythrocyte membrane to water is roughly tenfold greater than that of a phospholipid bilayer. Erythrocytes contain an integral membrane protein, called aquaporin, that acts as a water channel, allowing water to enter the erythrocyte cytoplasm. Frog eggs and oocytes apparently lack such a protein. Frog oocytes microinjected with mRNA encoding aquaporin swell in hypotonic solution, demonstrating that this protein is a water channel.

    For further study, see text section "Water Channels Are Necessary for Bulk Flow of Water across Cell Membranes," p. 610.

17. The high osmotic pressure inside plant cells results in an inward flow of water and outward turgor pressure against the plant-cell plasma membrane. If the same osmotic pressure could be experimentally induced inside mammalian cells, they would lyse. However, plant cells are surrounded by a rigid cell wall that is able to withstand the high turgor pressure from within the cell. Consequently, the plasma membrane does not lyse. If the cell wall were removed, the cell would lyse.

    For further study, see text section "Different Cells Have Various Mechanisms for Controlling Cell Volume," p. 609.

18a. The V_{max} and K_m values for glucose uptake can be estimated by inspection of the kinetic curves in Figure 15-1. The values are as follows:

Subject	K_m	V_{max}
DZ	6.0 mM	500 µmol glucose/ml-cells/h
KD	1.5 mM	100 µmol glucose/ml-cells/h
Control	1.5 mM	500 µmol glucose/ml-cells/h

18b. In the erythrocytes from patient DZ, the K_m for glucose uptake is about fourfold higher than in the controls. The higher K_m indicates a decreased affinity of the glucose uniporter for glucose, most likely due to an alteration in the glucose-binding site.

18c. In the erythrocytes from patient KD, the V_{max} for glucose uptake is about fivefold lower than in the controls. The lower V_{max} could be due to a fivefold decrease in the number of normal glucose uniporters present in the erythrocyte membrane. Alternatively, a normal number of defective glucose uniporters may be present. An alteration in uniporter amino acid sequence or in secondary modifications could lead to a decrease in V_{max}.

19a. The ouabain-binding site of the Na⁺/K⁺ ATPase must be located on the exoplasmic face of the plasma membrane, as microinjected ouabain does not inhibit the ATPase. The ATP-binding site is located on the cytoplasmic face of the plasma membrane, where it has access to ATP in the cytosol.

19b. The addition of a phosphate group to the enzyme would cause it to become more negatively charged. In the presence of extracellular ouabain, this negative charge, which normally is temporary, becomes permanent. Most likely, ouabain treatment inhibits the Na⁺/K⁺ ATPase by blocking the reaction at the point when phosphate has been added to the enzyme. Thus the phosphorylated form of the enzyme accumulates, blocking further ion transport. See MCB 4e, Figure 15-13 (p. 593).

19c. The observation that ouabain treatment results in the α subunit becoming more negatively charged but has no effect on the charge of the β subunit suggests that the α subunit is phosphorylated. This hypothesis could be tested by investigating the effect of a phosphatase on the charge properties of the α subunit following ouabain treatment.

20a. The two-Na⁺/one-glucose symporter is predominantly hydrophobic as shown by the hydrophobicity plot. As a very hydrophobic protein with little in the way of exposed non-membrane buried sequence, the protein is likely to be poorly antigenic. Consequently an antibody approach is not likely. Purifying the protein is difficult because it is not very abundant. Assuming a molecular weight of 65,000 daltons (~650 amino acids, each amino ~100 daltons), there are fewer than 20,000 copies of the protein per cell. Expression cloning is a good approach here.

20b. This is real world data. It is somewhat noisy. The exact prediction of transmembrane segment is therefore difficult and somewhat arbitrary. Transmembrane segments are typically amino acid stretches of about 20 hydrophobic amino acids in length. A reasonable estimate is 13 transmembrane domains.

20c. The prediction of 13 transmembrane domains is odd number and the N-terminal sequence itself is not hydrophobic. With the odd number of transmembrane domains, the N-terminus and C-terminus of the protein should be on opposite sides of the membrane. If the number of trans-membrane domains predicted were even, then the two ends of the protein would be on the same side of the membrane.

20d. The lysine side chain is positively charged and highly hydrophilic. Thermodyamically the placement of a positive charge in a hydrophobic lipid environment is unfavorable. Presumably, the positive charge is neutralized by association with a negatively charged amino acid residue in another transmembrane domain of the protein.

For further consideration of the structural properties of proteins, please see MCB 4e, Chapter 3.

21a. Assuming that no endogenous mouse L cell protein contribution is made to the tight junctions formed in the transfected cells, then the expression of the single protein claudin must be sufficient to generate tight junctions between cells.

21b. The tight junctions form over regions of cell-cell contact. The junctions do not go completely around the cell. Therefore, they cannot form an effective block to nutrient movement between cells within the cell layer or maintain separation between apical and basolateral cell surfaces.

21c. The tight junctions should consist of claudin-claudin protein complexes between neighboring cells. If the cell population were a mixed population of cells expressing either claudin tagged with epitope A (e.g., myc) or claudin tagged with epitope B (e.g., HA), then the claudin contribution of each cell could be distinguished by immunofluorescence using appropriate antibodies to the respective epitope tags. Note that antibodies to the core claudin protein sequence would not allow for distinctions to be made between claudins contributed by neighboring cells.

16

Cellular Energetics: Glycolysis, Aerobic Oxidation, and Photosynthesis

PART A: *Chapter Summary*

Cells use the energy released by hydrolysis of the terminal high-energy phosphoanhydride bond of ATP to power many otherwise energetically unfavorable processes. ATP is the most important molecule for capturing and transferring free energy in biological systems and can be thought of as the universal "currency" of chemical energy. Generation of ATP from ADP and P_I is an endergonic reaction (requiring an input of 7.3 kcal/mol) and results primarily from two main processes—aerobic oxidation and photosynthesis. In aerobic oxidation, oxygen and carbohydrates (principally glucose) are metabolized to CO_2 and H_2O and the released energy is used to synthesize ATP. The initial steps in the oxidation of glucose, called glycolysis, occur in the cytosol in both eukaryotes and prokaryotes and do not require O_2. The final steps, which require O_2, occur in mitochondria of eukaryotes and on the plasma membrane of prokaryotes. In photosynthesis, light energy is used to synthesize ATP and is also stored in the chemical bonds of carbohydrates. In plants and eukaryotic single-celled algae, photosynthesis occurs in chloroplasts; and several prokaryotes also carry out photosynthesis by a mechanism similar to that in chloroplasts. The O_2 generated during photosynthesis is the source of virtually all O_2 in the air, and the carbohydrates produced are the ultimate source of energy for virtually all nonphotosynthetic organisms.

Although aerobic oxidation and photosynthesis appear to have little in common, there are many striking similarities between these two processes. Bacteria, mitochondria, and chloroplasts all use the process of chemiosmosis to generate ATP from ADP and P_i. A transmembrane proton concentration gradient and electric potential (voltage gradient), termed the proton-motive force, are generated by the stepwise movement of electrons from higher to lower energy states via membrane-bound electron carriers. In mitochondria and nonphotosynthetic bacterial cells, electrons are transferred to O_2, the ultimate electron acceptor. In the thylakoid membrane of chloroplasts, energy absorbed from light removes electrons from water (forming O_2) and eventually these electrons are donated to CO_2 to synthesize carbohydrates. All these systems, however, couple electron transport to the pumping of protons to generate the proton-motive force. Moreover, all cells utilize essentially the same kind of membrane protein, an ATP synthase termed the F_0F_1 complex, to use the energy stored in the proton-motive force to synthesize ATP. Protons flow through the F_0F_1 complex from the exoplasmic to the cytosolic face of the membrane, driven by a combination of the proton-concentration gradient and the membrane-electric potential, leading to ATP formation on the cytosolic face of the membrane. Chemiosmotic coupling thus illustrates an important principle: the membrane potential; the concentration gradients of protons (and other ions) across a membrane; and the phosphoanhydride bonds in ATP are equivalent and interconvertible forms of chemical potential energy.

PART B: *Reviewing Concepts*

16.1 *Oxidation of Glucose and Fatty Acids to CO_2* (lecture date _____)

1. In the left-hand column of Table 16-1, the reactions involved in the breakdown of glucose to carbon dioxide are listed. Fill in the numbers showing the net production of the indicated molecules by each pathway listed on the left.

2. Write the chemical equation for the overall glycolytic pathway.

3. What is the similarity between lactic acid accumulation in mammalian muscle cells and ethanol accumulation in fermenting yeast cells?

4. The standard free-energy change $\Delta G^{o\prime}$ of many individual reactions in glycolysis is positive, indicating that these steps are not thermodynamically spontaneous. Why is it that glycolysis proceeds, nonetheless?

5. How are the protons and electrons produced during glycolysis important in generating energy for the cell, and what relationship do they have with mitochondria?

Overall reaction	Net production of			
	CO_2	ATP or GTP	NADH	$FADH_2$
1 glucose → 2 pyruvate	____	____	____	____
2 pyruvate → 2 acetyl CoA	____	____	____	____
2 acetyl CoA → 4 CO_2	____	____	____	____
glycolysis and citric acid cycle total	____	____	____	____

Table 16-1

16.2 *Electron Transport and Oxidative Phosphorylation* (lecture date _____)

6. The four mitochondrial electron-carrier complexes are listed in the left-hand column of Table 16-2. Indicate the order in which a pair of electrons from NADH or $FADH_2$ would move through these multiprotein complexes to O_2 by writing 1 (first), 2, or 3 (last) on the appropriate lines in the middle two columns. In the right-hand column, indicate how many protons are pumped by each carrier complex per electron pair.

7. Describe how ADP is transported into the mitochondrial matrix against a concentration gradient. What is the source of energy for this transport process?

8. Describe the experiment with chloroplast thylakoid membranes that demonstrated the role of a proton-gradient in ATP synthesis.

9. Each of the cytochromes in the mitochondria contain a heme prosthetic group, which permits these proteins to transport electrons in the electron transport system.

   a. Explain how heme prosthetic groups function as electron carriers.

   b. What property of the various cytochromes assures unidirectional electron flow along the electron transport chain?

   c. As electrons flow through the electron transport chain, they lose energy. How is much of this energy utilized?

Carrier Complex	Order of electron transfer		
	NADH $\rightarrow$ O_2	FADH$_2$ $\rightarrow$ O_2	No. protons pumped
CoQH$_2$-cytochrome c reductase	_____	_____	_____
NADH-CoQ reductase	_____	_____	_____
Cytochrome c oxidase	_____	_____	_____
Succinate-CoQ reductase	_____	_____	_____

Table 16-2

16.3 *Photosynthetic Stages and Light-Absorbing Pigments*
(lecture date _____)

10. Write the overall reaction for oxygen-generating photosynthesis.

11. Sometimes the phrase "dark reactions" is used to refer to carbon fixation. Why is this phrase somewhat inappropriate?

12. What complex contains the ATP-synthesizing enzyme of chloroplasts?

13. What does comparison of the absorption spectrum of chlorophyll a and the action spectrum of photosynthesis suggest about the function of chlorophyll a?

16.4 *Molecular Analysis of Photosystems*
(lecture date _____)

14. In the left-hand column of Table 16-3 are listed various components of the photosynthetic machinery in higher plants. During photosynthesis, these components are involved in linear and/or cyclic electron (e⁻) flow.

    a. During linear electron flow, electrons move from P$_{680}$ in PSII to NADP⁺, indicated by a 1 and 8 in the middle column of Table 16-3. Indicate the order in which electrons move through the remaining components of this pathway by writing in the appropriate numbers (2–7).

    b. During cyclic electron flow, electrons move from P$_{700}$ in PSI (1) through four other components and back to P$_{700}$ (6). In the right-hand column of Table 16-3, indicate the order in which electrons move through the other components of this pathway by writing in the appropriate numbers (2–5).

PSI/PSII component	Linear e⁻ flow	Cyclic e⁻ flow
NADP⁺	8	
P$_{700}$		1, 6
Cytochrome b/f		
P$_{680}$	1	
FAD		
Ferredoxin		
Quinone		
Plastocyanin		

Table 16-3

15. Under certain conditions, electron flow in photosynthetic purple bacteria is noncyclic. In this case, NAD⁺ is reduced, but O_2 is not evolved. How do purple bacteria accomplish this?

16. Why does O_2 accumulate on the luminal surface rather than the stromal surface of thylakoids?

17. Why does simultaneous illumination with 600-nm and 700-nm light support a higher photosynthetic rate in higher plants than the sum of the photosynthetic rates supported by illumination with light of each wavelength used separately?

16.5 CO₂ Metabolism during Photosynthesis
(lecture date _____)

18. In hot, dry environments, plants must keep their stomata closed much of the time to conserve water; as a result, the CO_2 level in the cells exposed to air falls below the K_m of ribulose 1,5-bisphosphate carboxylase for CO_2. What is the implication of this phenomenon? What special adaptation do some plants have that allow them to fix CO_2 under such conditions?

19. Where in the chloroplast do the reactions of the Calvin cycle occur? What is the major product of these reactions and how does that product reach the site of sucrose synthesis?

20. How is the sucrose produced as a result of photosynthesis delivered to the entire plant?

PART C: *Analyzing Experiments*

16.2 *Electron Transport and Oxidative Phosphorylation*

21. (••)A proton gradient can be analyzed with fluorescent dyes whose emission intensity profiles depend on pH. One of the most useful dyes for measuring the pH gradient across mitochondrial membranes is the membrane-impermeant, water-soluble fluorophore 2′,7′-bis-(2-carboxyethyl)-5(and 6)-carboxyfluorescein (BCECF). The effect of pH on the emission intensity of BCECF, excited at 505 nm, is shown in Figure 16-1. In one study, sealed vesicles containing this compound were prepared by mixing unsealed, isolated inner mitochondrial membranes with BCECF; after resealing the membranes, the vesicles were collected by centrifugation and then resuspended in nonfluorescent medium.

a. When these vesicles were incubated in a physiological buffer containing ADP, P_i, and O_2, the fluorescence of BCECF trapped inside gradually decreased in intensity. What does this decrease in fluorescent intensity suggest about this vesicular preparation?

b. After the vesicles were incubated in buffer containing ADP, P_i, and O_2 for a period of time, addition of dinitrophenol caused an increase in BCECF fluorescence. In contrast, addition of valinomycin produced only a small transient effect. Explain these findings.

c. When the vesicles (in the absence of uncouplers) were incubated with ATP rather than ADP and dilute hydrochloric acid was added to the buffer (on the outside of the vesicles), ATP hydrolysis was observed. Explain why this observation is predictable.

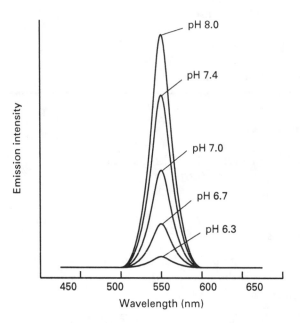

Figure 16-1

22. (••) The F_1 portion of the ATP synthases found in thylakoids are composed of five different polypeptides. These polypeptides, which can be distinguished from one another using SDS gel electrophoresis, are named α, β, γ, δ, and ε in order of their decreasing molecular weights. The subunit stoichiometry in the thylakoid F_1 particle is 3:3:1:1:1 with a particle molecular weight of approximately 400,000. It is known that the β subunits bind ADP and ATP and contain the catalytic sites, while the γ polypeptide acts as an inhibitor of ATP hydrolysis.

One experiment designed to improve our understanding of the thylakoid F_0F_1 particle subjected thylakoid membranes to three extraction methods (A, B, and C), with the goal of removing part(s) of the F_0F_1 particle in order to define the function of the extracted and unextracted portions. After extraction, the chloroplasts were exposed to a flash of light, which generated a pH gradient across the thylakoid membranes. In order to test the function of the F_0F_1 complex, the ability of the membrane preparations to dissipate the proton gradient was monitored. Figure 16-2a depicts the decrease in the pH gradient across unextracted thylakoid membranes and across those prepared with the three extraction techniques. These data show that method A produced membranes that dissipated the pH gradient at nearly the same rate as the no-extraction control. Method C produced membranes that were very leaky to protons, and method B produced membranes that dissipated the pH gradient at an intermediate rate.

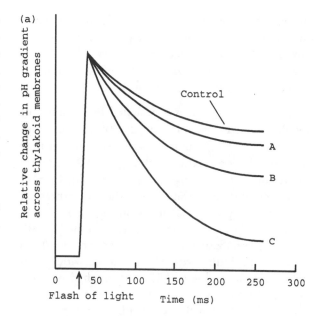

a. Suggest a method to determine which portion of the F_0F_1 complex was removed by each extraction method.

b. Immunoelectrophoresis of thylakoid membranes showed that methods A and B each extracted 13 percent of the F_1 from the membranes, while method C extracted 55 percent of the F_1. Because methods A and B had similar extraction efficiencies, but produced membranes that differed in proton permeability, the F_1 particles extracted by each method were partially purified and subjected to SDS gel electrophoresis. The results are shown in Figure 16-2b. What do these data suggest about the function of the δ subunit in the thylakoid F_0F_1 particle?

c. When the F_0 inhibitor N,N'-dicyclohexylcarbodiimide (DCCD) is added to membranes extracted by method C, these membranes exhibit the same profile for dissipation of the pH gradient as the unextracted control membranes. DCCD does not affect the extraction of the F_1 particle. Why is measurement of the proton flux in the presence of DCCD an important control in the investigation of the role of δ subunit in regulating proton flux through the thylakoid F_0F_1 particle?

d. Given the data suggesting that removal of the δ subunit results in membranes that are leaky to protons, propose more definitive experiments to test the hypothesis that this subunit acts as a "stopcock" of the F_0 portion.

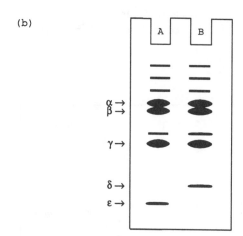

Figure 16-2

16.5 *CO₂ Metabolism during Photosynthesis*

23. (••) Ribulose 1,5-bisphosphate carboxylase (rubisco) catalyzes the initial reaction in carbon fixation, the addition of CO_2 to ribulose 1,5-bisphosphate (RuBP), as well as the competing reaction of O_2 with RuBP. In order for rubisco to be active, it must be carbamylated on a lysine residue by CO_2. In light, the enzyme rubisco activase activates rubisco by carbamylation; this reaction requires ATP hydrolysis. Activation of rubisco is strongly inhibited

by tight binding of RuBP to rubisco. RuBP binding to inactive rubisco induces an isomerization of the enzyme to a form in which RuBP is so tightly bound that its spontaneous release occurs with the extremely slow half-time of 35 min.

In one set of experiments designed to examine the regulation of rubisco activity, RuBP was added to rubisco so that tight binding occurred. Then at time 0, an incubation mixture with Mg^{2+}, ATP, and with or without rubisco activase was added. The carbon-fixing activity of rubisco was then measured at various time points, as shown in Figure 16-3a. In another experiment, the tightly bound RuBP-rubisco complex again was prepared; at time 0, an incubation mixture with Mg^{2+}, with or without rubisco activase, and with ATP or its nonhydrolyzable analog, ATP-γ-S was added to the complex. The release of RuBP from the RuBP-rubisco complex was monitored, as shown in Figure 16-3b.

a. Was rubisco activase necessary for activation of rubisco under the conditions used in these experiments?

b. Were rubisco activase, ATP hydrolysis, or both necessary for the dissociation of RuBP from rubisco?

c. Comparison of the kinetics of RuBP dissociation from rubisco and those of rubisco activation showed that 30 sec after exposure of the tightly bound RuBP-rubisco complex to rubisco activase, Mg^{2+}, and ATP, the specific activity of rubisco was 0.08 μmol CO_2 incorporated/min/mg; at this time, 65% of the RuBP had been released. At 150 sec, the specific activity was 0.79 μmol CO_2/min/mg; at this time, 98% of the RuBP had dissociated from the enzyme. What do these data suggest about the order in which dissociation of RuBP from rubisco and activation of rubisco took place?

d. Based on the data presented in this problem, summarize the events involved in the activation of inactive rubisco with tightly-bound RuBP, in the presence of Mg^{2+}, ATP, and rubisco activase.

(a)

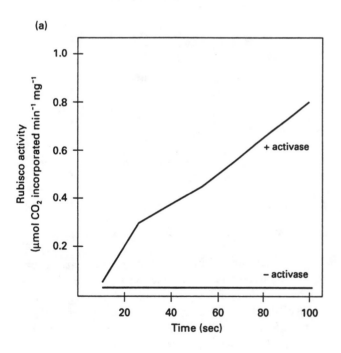

(b)

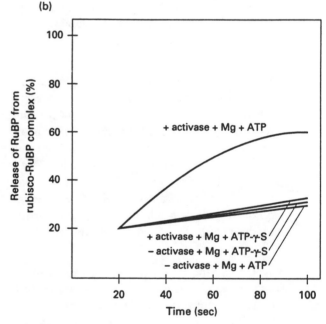

Figure 16-3

Part D: *Answers*

	Net production of			
Overall reaction	CO_2	ATP or GTP	NADH	$FADH_2$
1 glucose → 2 pyruvate	0	2 ATP	2	0
2 pyruvate → 2 acetyl CoA	2	0	2	0
2 acetyl CoA → 4 CO_2	4	2 GTP	6	2
glycolysis and citric acid cycle total	6	4	10	2

Table 16-4

1. See Table 16-4.

   For further study, see text Figures 16-3 (p. 620), 16-12 (p. 626), and Table 16-1 (p. 627).

2. The overall reaction for glycolysis is as follows:

$$C_6H_{12}O_6 + 2NAD^+ + 2ADP^{3-} + 2P_i^{2-} \rightarrow$$

$$2C_3H_4O_3 + 2NADH + 2ATP^{4-}$$

   For further study, see text section "Cytosolic Enzymes Convert Glucose to Pyruvate," p. 619.

3. Both lactic acid and ethanol accumulation results from the lack of O_2 in the environment. In anaerobic muscle cells, glucose cannot be broken down completely and lactic acid accumulates. In anaerobic yeast cells, fermentation of sugars results in the production of carbon dioxide and ethanol.

   For further study, see text section "Anaerobic Metabolism of Each Glucose Molecule Yields Only Two ADP Molecules," p. 619.

4. In the glycolytic pathway, endergonic reactions are coupled with exergonic reactions, the latter pulling glycolysis forward. As a result, the $\Delta G^{\circ\prime}$ for the entire pathway is negative.

   For further study, see text section "Substrate-Level Phosphorylation Generates ATP during Glycolysis," p. 619.

5. The four electrons and two of the four protons produced in glycolysis are used to reduce the electron carrier NAD^+ (nicotinamide adenine dinucleotide) to NADH. The electrons of NADH are transferred via an electron shuttle to the mitochondrial matrix, from which they can enter the electron transport chain in the inner mitochondrial membrane. Thus the energy of these electrons is used to produce the proton-motive force and is ultimately converted into ATP by the F_0F_1 ATP synthase.

   For further study, see text section "Inner-Membrane Proteins Allow the Uptake of Electrons from Cytosolic NADH," p. 626.

6. See Table 16-5.

Table 16-5

Carrier complex	Order of electron transfer		No. protons pumped
	NADH $\rightarrow$ O_2	$FADH_2 \rightarrow$ O_2	
CoQH$_2$-cytochrome c reductase	2	2	4
NADH-CoQ reductase	1		4
Cytochrome c oxidase	3	3	2
Succinate-CoQ reductase		1	0

For further study, see text section "Electrons Flow from $FADH_2$ and NADH to O_2 via a Series of Multiprotein Complexes," p. 634.

7. ADP^{3-} is transported into the mitochondrial matrix while ATP$_{4-}$ is transported outward by the ATP-ADP antiport protein. At the same time, P$_i^{2-}$ is transported inward and OH$^-$ is transported outward by the phosphate antiporter. The OH$^-$ combines with a H$^+$ in the intermembrane space using the energy of the proton-motive force to drive the combined processes. It also can be noted that ATP^{4-}-ADP^{3-} exchange is favored in terms of the membrane electrical gradient, as a net negative charge is moved outward, down the electrical gradient.

For further study, see text section "Transporters in the Inner Mitochondrial Membrane Are Powered by the Proton-Motive Force," p. 646.

8. Isolated chloroplast thylakoid vesicles were equilibrated in the dark with a solution at pH 4.0 to allow the pH of the thylakoid lumen to reach 4.0. The vesicles were then rapidly exchanged into a solution at pH 8.0 that contained ADP and P$_i$. The resulting proton gradient (high in the lumen, low in the surrounding solution) caused protons to move through the F$_0$F$_1$ complexes in the membrane and provided the energy to synthesize ATP from the ADP and P$_i$ in the surrounding solution. This experiment thus provided direct evidence that a chemiosmotic mechanism could power ATP synthesis.

For further study, see text section "Experiments with Membrane Vesicles Support the Chemiosmotic Mechanism of ATP Formation," p. 641.

9a. Each of the heme prosthetic groups present in cytochromes contains an iron atom that accepts an electron as it is reduced and releases an electron as it is oxidized. Because the heme ring has numerous resonance forms, the second electron is delocalized to the heme carbon and nitrogen atoms.

9b. The various cytochromes in the electron transport chain contain heme prosthetic groups with different axial ligands [see text Figure 16-21 (p. 637)]. As a result, each cytochrome has a different reduction potential, so that electrons can move only in a single order through the electron carriers.

9c. Much of the energy lost by electrons moving through the electron transport chain is used to pump protons from the matrix to the intermembrane space, thus generating the proton-motive force. See text Figure 16-19 (p. 636).

For further study, see text section "Electrons Flow from $FADH_2$ and NADH to O_2 via a Series of Multiprotein Complexes," p. 634.

10. $6CO_2 + 6H_2O \rightarrow 6O_2 + C_6H_{12}O_6$

For further study, see text section "Photosynthetic Stages and Light-Absorbing Pigments," p. 648.

11. The Calvin cycle reactions that result in carbon fixation can occur both in dark and light conditions. However, these reactions, which depend on the light-dependent reactions to provide ATP and NADPH, are inhibited by dark conditions. The activity of certain of these dark reaction enzymes is lower in the dark than in the light for several reasons: a decrease in the pH and Mg^{2+} concentration of the stroma in the dark; oxidation of the stromal enzymes in the dark; and a lack of activation of ribulose 1,5-bisphosphate carboxylase by rubisco activase under dark conditions.

For further study, see text section "Carbon Fixation," p. 651.

12. The thylakoid F_0F_1 complex contains the ATP synthase in chloroplasts.

For further study, see text section "Generation of ATP," p. 651.

13. The absorption spectrum of chlorophyll *a* is similar but not identical to the action spectrum of photosynthesis (i.e., the relative rate of photosynthesis at various wavelengths of incident light). This suggests that chlorophyll *a* is critical to photosynthesis, but that other pigments also are involved. In fact, photosynthesis driven by light of 500 nm is primarily due to absorption by carotenoids and chlorophyll *b*, whereas photosynthesis driven by light of 680 nm is primarily due to absorption by chlorophyll *a*.

For further study, see text section "Chlorophyll *a* is Present in Both Components of a Photosystem," p. 651.

14. See Table 16-6.

PSI/PSII component	Linear e⁻ flow	Cyclic e⁻ flow
$NADP^+$	8	
P_{700}	5	1, 6
Cytochrome *b/f*	3	4
P_{680}	1	
FAD	7	
Ferredoxin	6	2
Quinone	2	3
Plastocyanin	4	5

Table 16-6

For further study, see text section "Chloroplasts Contain Two Functionally and Spatially Distinct Photosystems," p. 658; and "Cyclic Electron Flow in PSI Generates ATP but No NADPH," p. 661.

15. During noncyclic electron flow in purple bacteria, the loss of an electron by the reaction center causes cytochrome *c* to donate an electron to the reaction center. Hydrogen gas (H_2) or hydrogen sulfide (H_2S) rather than H_2O donates electrons to reduce oxidized cytochrome *c*. The electron lost from the reaction center is eventually used to reduce NAD^+ to NADH. The reducing power of NADH is used to fix CO_2.

For further study, see text section "Noncyclic Electron Transport," p. 657.

16. Absorption of light causes electrons to move from P_{680} to an acceptor quinone on the stromal surface, thus creating a transient positive charge on the luminal side of the thylakoid membrane, where the reaction-center chlorophyll is located. As a consequence, H_2O is split by the oxygen-evolving complex on the luminal surface, electrons are transferred to the oxidized reaction-center chlorophyll, and O_2 and protons are released into the thylakoid lumen.

For further study, see text Figures 16-38 (p. 653) and 16-42 (p. 658).

17. The two photosystems in higher plants absorb light of differing wavelengths. PSII absorbs light of 680 nm, and can use only light of this wavelength or shorter, whereas PSI absorbs light of 700 nm, and can use only light of this wavelength or shorter. Alone, light of wavelengths greater than 680 nm is not efficient in supporting photosynthesis, since the chlorophyll a in PSII cannot be excited by such light. However, when light of shorter wavelength is supplied to raise the energy of electrons generated by PSII, the 700-nm light can be used by PSI and photosynthesis is enhanced over that supported by 600-nm light alone. This phenomenon is called the Emerson effect.

For further study, see text section "Chloroplasts Contain Two Functionally Distinct Photosystems," p. 658.

18. If the level of CO_2 falls below the K_m of ribulose 1,5-bisphosphate carboxylase, photorespiration, in which ribulose 1,5-bisphosphate carboxylase utilizes O_2 rather than CO_2 as a substrate, is favored over carbon fixation. Photorespiration is a wasteful process that consumes ATP. To avoid this problem, plants such as corn, cane sugar, and crabgrass, termed C_4 plants, have developed a two-step mechanism of CO_2 fixation in which the initial assimilation of CO_2 can occur in a relatively low CO_2, high O_2 environment. The leaves of C_4 plants possess two types of chloroplast-containing cells: mesophyll cells, which are directly exposed to air, and bundle sheath cells, which underlie the mesophyll cells. (The leaves of other plants, termed C_3 plants, possess chloroplast-containing mesophyll cells but lack bundle sheath cells.) In C_4 plants, the bundle sheath cells generally contain more chloroplasts (and hence Calvin cycle enzymes) than the mesophyll cells. In the mesophyll cells, CO_2 reacts with phosphoenolpyruvate to produce the four-carbon compound oxaloacetate, which is reduced to malate. The enzyme catalyzing this reaction is active even at low CO_2 levels. Malate is shuttled to the bundle sheath cells, where it is decarboxylated, releasing CO_2 and thereby producing a relatively high CO_2, low O_2 environment. Under these conditions, ribulose 1,5-bisphosphate carboxylase operates to fix CO_2 and the Calvin cycle can operate as usual.

For further study, see text section "Photorespiration, Which Consumes O_2 and Liberates CO_2, Competes with Photosynthesis," p. 667; and "The C4 Pathway for CO_2 Fixation is Used by Many Tropical Plants," p. 667.

19. The reactions of the Calvin cycle take place in the stromal space of the chloroplast. The major product of these reactions is glyceraldehyde 3-phosphate, which is transported to the cytosol for use in sucrose synthesis. Transport of glyceraldehyde 3-phosphate from the stromal space into the cytosol is carried out by the phosphate-triosephosphate antiport protein, which simultaneously transports P_i from the cytosol into the stromal space.

For further study, see text section "CO_2 Fixation Occurs in the Chloroplast Stroma," p. 664, and "Synthesis of Sucrose Incorporating Fixed CO_2 Is Completed in the Cytosol," p. 665.

20. Sucrose, which is produced in photosynthetic cells, is delivered to other cells via the cellular phloem. The phloem, which is composed of sieve-tube cells and companion cells, forms a continuous tube of cytosol throughout the entire plant. Sucrose moves through the phloem due to forces generated by osmotic pressure differences.

For further study, see text section "Sucrose is Transported from Leaves through the Phloem to All Plant Tissues," p. 670.

21a. The electron transport system normally pumps protons out of the mitochondrial matrix, increasing the pH of the matrix; thus the fluorescence of matrix-trapped BCECF would increase in intensity. The observed decrease in intensity of BCECF trapped inside the vesicles suggests that the vesicles have an inverted (inside-out) orientation, so that protons were pumped from the outside to the inside of the vesicles.

21b. Dinitrophenol compromises the pH gradient and the resulting equilibration of protons leads to an increase in the intravesicular pH and corresponding increase in emission intensity. Valinomycin, a potassium ionophore, affects the electric potential more than the pH gradient. Since BCECF fluorescence reflects the pH of the milieu, it is largely unaffected by valinomycin-induced changes in the transmembrane electric potential.

21c. The phosphorylation of ADP by ATP synthase on the inner mitochondrial membrane is coupled to and depends on the movement of protons down the pH gradient from the outside to the inside of the mitochondria. Exposure of the inside-out vesicles to dilute acid reverses the normal pH gradient; as a result, the ATP synthase hydrolyzes ATP to ADP.

For further study, see text section "Experiments with Membrane Vesicles Support the Chemiosmotic Mechanism of ATP Formation," p. 641.

22a. The membranes could be separated from the extracted material by centrifugation. SDS gel electrophoresis of the extracted material in the supernatant and of the pelleted membranes would reveal which subunits of the complex were present in the membranes and which were extracted. Immunoblotting of the gel could be used to verify the identification of the bands, corresponding to the ATP synthase subunits.

22b. The material extracted by method B contains the δ subunit, whereas that extracted by method A does not. Thus the membranes left after extraction by method B are deficient in the δ subunit. The finding that thylakoid membranes extracted by method B, and thus deficient in δ, are leaky suggests that the δ subunit blocks proton flow through the membrane-bound F_0 particle.

22c. Addition of DCCD, which is known to interact with F_0, allows one to distinguish proton movement through F_0 from proton movement through other integral membrane proteins or parts of the (leaky) membrane. A full block (i.e., DCCD-dependent return to control levels) indicates that the observed loss of the pH gradient upon extraction by method C is probably all mediated through the F_0 particle.

22d. This hypothesis could be tested by isolating the δ subunit by chromatography and then adding it to the membranes resulting from extraction

method B. If δ acts as a stopcock, then the pH gradient profile of membranes extracted by method B with added δ subunit should be the same as that of unextracted membranes. Another approach would be to examine the function of the F_0F_1 complex in thylakoid membranes isolated from plants with nonlethal mutations in the gene encoding the δ subunit. The finding that changes in the amino acid sequence of the δ subunit alters the permeability of the thylakoid membrane to protons would support the hypothesis.

For further study, see text section "ATP Synthase Comprises a Proton Channel (F_0) and ATPase (F_1)," p. 664, and "The F_0F_1 Complex Harnesses the Proton-Motive Force to Power ATP Synthesis," p. 645.

23a. The data in Figure 16-3a show that CO_2 fixation by rubisco requires activation of the enzyme by rubisco activase.

23b. As shown in Figure 16-3b, both were probably necessary. Dissociation appears to have been greater in the presence of ATP than in the presence of its nonhydrolyzable analog, suggesting that hydrolysis is necessary. An alternative explanation of the data is that dissociation requires binding of ATP to rubisco activase but not its hydrolysis and that ATP-γ-S did not bind to rubisco activase. However, other experimental evidence indicates that the analog does bind, so ATP hydrolysis most likely is required.

23c. These data suggest that dissociation of RuBP from rubisco occurred before activation of the enzyme. At 30 sec, dissociation was already 65% complete, whereas the rubisco CO_2 fixation activity was only about 10% of the activity at 150 sec.

23d. Rubisco activase catalyzes the dissociation of tightly bound RuBP from rubisco. This dissociation requires ATP hydrolysis. Dissociation is followed by activation of rubisco in a reaction also catalyzed by rubisco activase. Data not presented in this problem have demonstrated that this latter reaction requires ATP hydrolysis too.

For further study, see text section "CO_2 Metabolism during Photosynthesis," p. 664.

17

Protein Sorting: Organelle Biogenesis and Protein Secretion

PART A: *Chapter Summary*

The typical mammalian cell contains up to 10,000 kinds of proteins and a yeast cell about 5,000. For normal cell function, all of these proteins must be localized to the proper subcellular compartment. The Na^+, K^+ ATPase pump must, for example, be distributed to the basolateral surface of the intestinal epithelial cell. Water-soluble components, such as RNA and DNA must be targeted to the nucleus; still other proteins must be delivered to the endoplasmic reticulum and Golgi apparatus. Many proteins, such as hormones and components of the extra-cellular matrix, must be directed to the cell surface and secreted.

The process of directing each newly made polypeptide to a particular destination—referred to as protein-targeting, or sorting—is essential to the organization and functioning of eukaryotic cells. The process occurs at several levels. A small number of proteins, encoded by DNA present in mitochondria and chloroplasts, are synthesized on ribosomes in these organelles and incorporated directly into compartments within these organelles. Most mitochondrial and chloroplast proteins and all proteins of the other organelles, particles, and membranes of a eukaryotic cell are encoded by nuclear DNA. They are synthesized on ribosomes in the cytosol and distributed to their correct destinations by the sequential action of up to several sorting signals. How nuclear-encoded organelle, membrane, and secretory proteins are sorted to their correct destinations is the major subject of this chapter.

The first sorting event occurs during initial growth of nascent polypeptide chains on cytosolic ribosomes. Some nascent chains contain, generally at their amino terminus, a specific signal, or targeting-sequence that directs the ribosomes, synthesizing them to the endoplasmic reticulum (ER). Protein synthesis is completed in association with membranes of the rough ER. The completed polypeptide chains may then either stay in the ER, or move to the Golgi apparatus and subsequent sorting to other possible destinations. Proteins synthesized and sorted in this pathway, termed the secretory pathway, include not only those that are secreted from the cell, but also enzymes and other resident proteins in the lumen of the ER, Golgi apparatus, and lysosomes, as well as integral proteins in the membranes of these organelles and the plasma membrane.

Synthesis of all other nuclear-encoded proteins is completed on free non-membrane attached cytosolic ribosomes, and the completed proteins are released into the cytosol. These proteins remain in the cytosol unless they contain a specific signal sequence that directs them to the mitochondrion, chloroplast, peroxisome, or nucleus. Many of these proteins are subsequently sorted further to reach their correct destinations within these organelles; such sorting events depend on multiple signals within the protein. Each sorting event involves binding of a signal sequence to one or more receptor proteins on the surface or interior of the organelle.

This chapter details the mechanisms whereby proteins are sorted to the major organelles and compartments of the cell. (The transport of proteins in and out of the nucleus was described earlier, in Chapter 11, Section 11.4) We first cover targeting of proteins to mitochondria, chloroplasts, and peroxisomes. The next several sections describe the various components and events in the secretory pathway, including the post-translational modifications that occur to proteins as they move through this pathway. We then discuss how the proteins are internalized into cells following binding to specific cell-surface receptors and the fate of such internalized proteins. Finally, we concentrate on the machinery associated with various small membrane-bounded vesicles that carry proteins within cells and deliver their contents to specific destinations.

PART B: *Reviewing Concepts*

17.1 *Synthesis and Targeting of Mito-chondrial and Chloroplast Proteins*
(lecture date _____)

1. Antibodies specific for the C-terminal end of a cytosol-synthesized mitochondrial protein can prevent completion of translocation of the precursor protein into the mitochondrial matrix, although a portion of the precursor is translocated. Explain these observations.

2. What is the salient difference between the energy requirement for translocation of proteins to mitochondria and to chloroplasts?

17.2 *Synthesis and Targeting of Peroxi-somal Proteins* (lecture date _____)

3. Fibroblasts isolated from patients with Zellweger syndrome are incapable of translocating catalase and other lumenal proteins synthesized in the cytosol into peroxisomes. How could you demonstrate that Zellweger fibroblasts can synthesize catalase?

4. One hypothesis that has been advanced to explain the nature and origin of the peroxisome proposes that this organelle is a vestige of the organellar site that arose in primitive, premitochondrial cells to handle oxygen introduced into the atmosphere by photosynthetic bacteria. Since oxygen radicals and some oxygen-containing molecules can be highly toxic to cells, a mechanism for handling them would be necessary for cell survival. How consistent is this hypothesis with the structure and function of peroxisomes?

17.3 *Overview of the Secretory Pathway*
(lecture date _____)

5. Describe the overall process by which a cell-surface glycoprotein such as glycophorin is synthesized and transported to the cell surface.

6. Cisternal progression is the commonly accepted model today for how transport through the Golgi apparatus occurs. What is the role(s) of Golgi associated vesicles in the cisternal progression model?

17.4 *Translocation of Secretory Proteins across the ER Membrane*
(lecture date _____)

7. The signal recognition particle (SRP) is involved in regulating the elongation of nascent secretory proteins and targeting them to the endoplasmic reticulum. Describe an experiment in which these functions of SRP have been demonstrated.

8. Energy input is required for protein translocation across the ER membrane. What is the source of this energy in yeast and mammalian cells?

17.5 *Insertion of Membrane Proteins into the ER Membrane* (lecture date
_____)

9. In some proteins, a signal-sequence also functions as a topogenic sequence. Discuss how these dual-function signal sequences differ from monofunctional signal sequences, which simply direct nascent polypeptides to the ER membrane.

10. Describe the process by which a protein becomes glycosylphosphatidylinositol (GPI) anchored?

17.6 *Post-Translational Modifications and Quality Control in the Rough ER*
(lecture date _____)

11. Processing of proteins is subject to quality-control within the ER, which prevents improperly folded proteins from reaching the cell surface. What are typical fates of the misfolded proteins?

12. What is the expected subcellular distribution of protein disulfide isomerase (PDI) lacking a C-terminal KDEL sequence?

17.7 *Protein Glycosylation in the ER and Golgi Complex* (lecture date _____)

13. What are the protein amino acids modified during N-linked and O-linked glycosylation?

14. Explain why primate cells, rather than non-primate cells or bacteria, are the preferred hosts for insertion of cDNA for the production of human secretory proteins of commercial value.

15. When fibroblasts from patients suffering from I cell disease are cultured, they secrete lysosomal enzymes rather than accumulating them within lysosomes. When lysosomal enzymes containing M6P are added to the culture, they are rapidly internalized and accumulate within lysosomes. Why don't the I cell secreted lysosomal enzymes accumulate within cells?

17.8 *Golgi and Post-Golgi Protein Sorting and Proteolytic Processing* (lecture date _____)

16. How does regulated secretion differ from constitutive secretion?

17. Specialization of membranes is a characteristic of mammalian cells. What are the mechanisms by which the apical and basolateral surfaces of certain cells become differentiated?

17.9 *Receptor-Mediated Endocytosis and the Sorting of Internalized Proteins* (lecture date _____)

18 Many extracellular proteins are internalized by receptor-mediated endocytosis. What are the molecular signals that trigger uptake of a protein by receptor-mediated endocytosis?

19. The pH of a compartment can be critical for association or dissociation of receptor and ligand. How does the acidic pH of the late endosome result in different fates for the LDL and transferrin receptors?

17.10 *Molecular Mechanisms of Vesicular Traffic* (lecture date _____)

20. What are the known types of coat-proteins and which small GTPase is required for the transport of each type to the membrane?

21. What is the role of SNARE proteins in vesicle fusion?

PART C: *Analyzing Experiments*

17.1 *Synthesis and Targeting of Mitochondrial and Chloroplast Proteins*

22. (•••) The light-harvesting complex protein (LHCP) is an example of a chloroplast protein which is synthesized in the cytosol and then imported and subsequently processed to a mature form. It is found in thylakoid membranes of the chloroplast.

In an experiment to understand the forces governing translocation of preLHCP, the *in vitro* translocation of [^{3}H]preLHCP to isolated chloroplasts was measured. After an appropriate time of incubation, the chloroplasts were lysed and a fraction containing both thylakoid and envelope proteins was prepared and analyzed by SDS-PAGE autoradiography, as depicted in lane 2 of Figure 17-1. Lane 1 is a control incubation of labeled preLHCP in the absence of chloroplasts. Following incubation, samples of the thylakoid membranes were treated with NaOH or a protease and then analyzed, as depicted in lane 3 (NaOH treatment), lane 4 (thermolysin treatment), and lane 5 (trypsin treatment).

a. What does the difference in the migration patterns in lanes 1 and 2 in Figure 17-3 indicate about preLHCP? Why is this a critical part of the experiment?

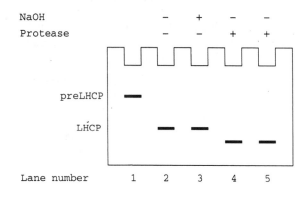

Figure 17-1

b. Why were NaOH and proteases used in this experiment?

c. In Figure 17-1, what does the difference in the migration patterns in lanes 4 and 5, compared with the pattern in lane 2, suggest about the location of LHCP in the thylakoid membrane?

d. In other experiments, purified thylakoid membranes prepared on a sucrose gradient, were incubated with labeled preLHCP in the presence and absence of ATP, stroma, and protease. After an appropriate time, the samples were analyzed by SDS-PAGE autoradiography. The resulting gel profiles are depicted in Figure 17-2. What conclusions can be drawn from these data about the effects ATP and stroma have on binding of preLHCP to thylakoid membranes?

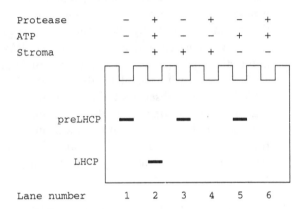

Figure 17-2

17.4 Translocation of Secretory Proteins across the ER Membrane

23. (•) As illlustrated in Figure 17-3, protein synthesis may be done in the presence or absence of microsomal membranes. Under what conditions will protein translocate into the microsomal membranes, and why?

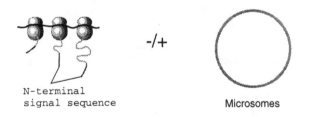

Figure 17-3

17.4 Translocation of Secretory Proteins across the ER Membrane

24. (•••) The signal recognition particle (SRP) consists of six polypeptide subunits (9, 14, 19, 54, 68, and 72 kDa), which are organized into three different functional entities surrounding a 7S-RNA molecule. After protein synthesis is initiated in the cytoplasm, the 54-kDa subunit of the SRP binds to the signal sequence shortly after it is synthesized on the ribosome. This SRP-ribosome unit then associates with the SRP receptor (docking protein) on the endoplasmic reticulum where synthesis continues and the newly synthesized protein is translocated across the endoplasmic reticulum.

In experiments designed to sort out the factors necessary to promote translocation, a complete cell-free translational system was incubated with a mRNA encoding a secretory protein of 40-kDa when fully modified, [35S]methionine to monitor protein synthesis, and various preparations containing different factors, as indicated in Table 17-1 (next page). GMP-PNP and AMP-PNP are non-hydrolyzable analogs of GTP and ATP, respectively. After each of the preparations was incubated with the translational system in appropriate buffers, the sample was incubated with a protease (proteinase K). All proteins were then precipitated, denatured, and separated on an SDS gel. Autoradiography of the gel revealed the pattern shown in Figure 17-4. Each lane of the autoradiogram is labeled with the corresponding preparation number.

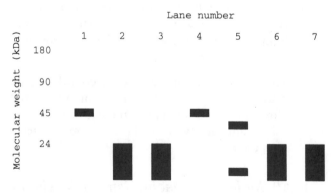

Figure 17-4

a. What is the significance of the discrete bands in lanes 1, 4, and 5 of the autoradiogram and of the diffuse bands in the other lanes?

b. What can you conclude from this experiment about the factors required for protein translocation and the mechanisms involved in this process?

Preparation number	Components
1	SRP-ribosome complex + microsomes + ATP + GTP
2	SRP-ribosome complex + microsomes + ATP + GTP + 5mM EDTA
3	SRP-ribosome complex + microsomes + AMP-PNP + GTP
4	SRP-ribosome complex + microsomes + ATP + GMP-PNP
5	SRP-ribosome complex + microsomes + ATP + GTP + more time for incubation
6	SRP-ribosome complex + microsomes
7	SRP-ribosome complex + microsomes + ATP

Table 17-1

17.5 *Insertion of Membrane Proteins into the ER Membrane*

25. (●●) Figure 17-5 schematically depicts proteins containing various types of signal and topogenic sequences. Predict the arrangement of each type of protein shown in the Figure with respect to the endoplasmic reticulum membrane and lumen.

17.10 *Molecular Mechanisms of Vesicular Traffic*

26. (●●●) Type I transmembrane proteins are localized to the ER of mammalian cells by virtue of a common signal consisting of two lysine residues at positions −3 and −4 from the C-terminal end of the cytoplasmic domain. A similar dilysine signal has been recently identified in the cytoplasmic domain of a yeast ER membrane protein. In both mam-

malian and yeast cells, this dilysine sequence appears to function as a retrieval sequence; analysis of the posttranslational modifications of the proteins indicates that they have been retrieved from the Golgi apparatus.

Genetic approaches have been very important to investigations of the machinery involved in the retrieval of type I transmembrane proteins to the ER. In such experiments, a series of yeast temperature-sensitive mutants defective in retrieval were isolated and then characterized for the nature of the defect. The nine temperature-sensitive mutations isolated fell into two complementation classes: *ret1* for retrieval defective 1 (eight examples all in one allele) and *sec21* (one example, which was a new *sec21* allele termed *sec21-2*). The *sec21* gene was originally identified on the basis of a temperature-sensitive defect in secretion (*sec21-1* allele).

To quantify the effect of mutations on retrieval, the fate of a fusion protein containing the dilysine (diK) retrieval sequence was determined. If this diK-fusion protein is retrieved, it is not exposed to a post-ER protease. After the fusion protein was expressed in yeast cells incubated with a radiolabeled amino acid, the labeled products were resolved by gel electrophoresis and detected by autoradiography. Figure 17-6a (next page) depicts the resulting autoradiograms for wild-type (WT) yeast cells and the *ret1-1, sec21-1,* and *sec21-2* mutants. In the absence of retrieval, the diK-fusion protein is processed by the post-ER protease.

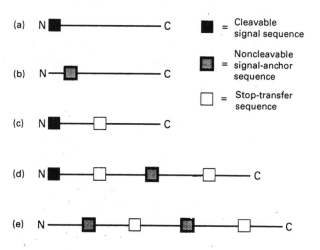

Figure 17-5

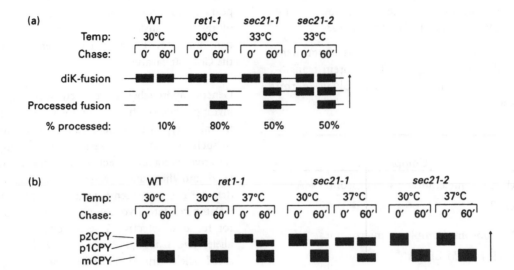

Figure 17-6

To quantify the effect of these mutations on secretion, researchers analyzed processing of carboxypeptidase Y (CPY) by a similar approach. CPY exists in two precursor forms termed p1CPY and p2CPY, which are present in the ER and Golgi complex, respectively; the mature form of this enzyme (mCPY) is present in the vacuole. The autoradiograms from this analysis are represented in Figure 17-6b.

a. Based on the data in Figure 17-6a, do all three mutations have a similar effect on processing of the diK-fusion protein?

b. Based on the data in Figure 17-6b, do all three mutations have a similar effect on secretion of CPY to the vacuole?

c. What evidence regarding multiple functional domains in the Ret1 protein and Sec21 protein do these results provide?

d. Surprisingly Ret1p is a subunit of the yeast COP coatomer. In *Saccharomyces cerevisiae*, β-COP, β'-COP, and γ-COP are encoded by the *sec26*, *sec27*, and *sec21* genes, respectively. Mutations in *sec26* and *sec27* also affect retrieval of the diK–fusion protein. What do these data suggest regarding the role of COP proteins in retrieval and secretion?

27. (••) Donor and acceptor membranes can be used in a test tube assay to establish the biochemical requirements for protein transfer from one Golgi subcompartment to another. In a typical assay, the donor membranes are isolated from cells with a mutation in a specific glycosyltransferase, commonly N-acetylglucosamine transferase-1, which is located in the *medial* Golgi. Mutant cells that are infected with vesicular stomatitis virus (VSV) are unable to process completely the virus-encoded G protein, a glycoprotein, *en route* to the cell surface. Incompletely processed sugar side chains carried by solubilized G protein are sensitive to digestion by endoglycosidase H. Treatment of the sensitive form of the glycoprotein (G_s) with this enzyme yields a product that migrates more rapidly in a polyacrylamide gel than does the resistant form (G_R), thus providing an assay for the two forms.

In one experiment, VSV G protein in mutant donor cells was prelabeled by a brief (pulse) exposure of the cells to [³H]palmitate. VSV G protein is acylated with palmitate in the *cis*-Golgi complex. The cells were then either homogenized immediately or after various chase times in isotope-free media. The donor membranes from the homogenates then were incubated with acceptor membranes from wild-type cells; after 30 min, G protein was solubilized from the *in vitro* membrane mixtures and assayed for sensitivity to endoglycosidase H. Figure 17-7a (next page) illustrates autoradiograms from this assay for two *in vivo* chase-times. Figure 17-7b (next page) shows the quantitative effect of chase-time on the extent of processing of G protein *in vitro*.

a. In this experimental protocol, what results would indicate that protein transport between *cis* - and the *media*-Golgi has occurred?

b. What is the effect of chase-time on the sensitivity of VSV G protein to endoglycosidase H? How does this effect relate to protein transport between Golgi subcompartments?

c. What evidence does this experiment provide regarding whether cargo protein transport such as that of VSV-G protein through the Golgi apparatus is unidirectional?

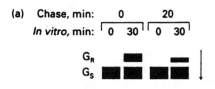

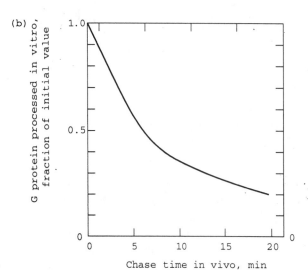

Figure 17-7

Answers

1. Because the precursor protein contains an N-terminal matrix-targeting signal, the N-terminus of the precursor molecule is translocated first. An antibody to the C-terminal end would not prevent passage of the N-terminal end across the mitochondrial membrane. However, the large size of the antibody bound to the C-terminal end would inhibit passage of the entire protein into the organelle, resulting in a translocation intermediate. Such intermediates have been generated in other ways, for example, by the addition of methotrexate during the translocation of dihydrofolate reductase. In the presence of the drug, which binds to the protein in its native configuration, translocation is blocked. Experi-ments such as these indicate that a protein must be in an unfolded state to pass into the interior of the mitochondrion. See MCB, Figure 17-5 (p. 683).

For further study, see text section "Studies with Chimeric Proteins Confirm Major Features of Mitochondrial Import," p. 682.

2. Energy is required for translocation of proteins to both chloroplasts and mitochondria. Three separate inputs of energy are required for mitochondrial import: ATP hydrolysis in the cytosol, a proton-motive force across the inner mitochondrial membrane, and ATP hydrolysis in the mitochondrial matrix. Protein imported into the chloroplast

strom appears to be powered solely by ATP hydrolysis. ATP hydrolysis in all cases appears to be related to the chaperone proteins, be they cytosolic or present in the mitochondrial matrix or chloroplast stroma.

For further study, see text sections "The Uptake of Mitochondrial Proteins Requires Energy," p. 682 and "Targeting to the Chloroplast Stromal Space," p. 686.

3. The ability of Zellweger fibroblasts to synthesize catalase could be demonstrated by carrying out protein synthesis with a cytosolic extract and then adding antibodies to catalase protein to immunoprecipitate newly-synthesized catalase in the reaction mixtures.

For further study, see text section "Peroxisomal Protein Import Is Defective in Some Genetic Diseases," p. 690.

4. Peroxisomes contain enzymes that degrade amino acids and fatty acids, generating hydrogen peroxide (H_2O_2), which is potentially toxic to cells. The peroxisomal enzyme catalase decomposes H_2O_2 into H_2O. In view of this, the hypothesis that peroxisomes originated to cope with oxygen in primitive cells is reasonable.

For further study, see text section "Synthesis and Targeting of Peroxisomal Proteins," p. 689.

5. A membrane protein such as glycophorin is synthesized by ribosomes that associate with the rough endoplasmic reticulum. From there, the protein is transported via vesicles to the *cis* Golgi reticulum (network). In a process termed *cisternal progression* or *maturation*, the *cis*-Golgi reticulum matures into the *cis*-Golgi, *medial*- and *trans*-Golgi. Progression completes with maturation of the Golgi cisterna into the *trans*-Golgi reticulum (network). From the *trans*-Golgi reticulum, the portein is transported via a vesicular carrier to the plasma membrane. During the whole process the topology of the protein with respect to the membrane is maintained.

For further study, see text section "Secretory Proteins Move from the Rough ER Lumen through

the Golgi Complex and Then to the Golgi Apparatus," p. 692.

6. In the cisternal progression model, small retrograde vesicular carriers retrieve membrane and luminal proteins from later to earlier Golgi cisternae. By this process, enzymes and other Golgi resident proteins come to be localized either in the *cis*- or *medial*- or *trans*-Golgi cisternae.

For further study, see text section "Secretory Proteins Move from the Rough ER Lumen through the Golgi Complex and Then to the Golgi Apparatus," p. 692.

7. The functions of SRP were demonstrated in a series of experiments utilizing a cell-free protein-synthesizing system and mRNA encoding pre-prolactin, a typical secretory protein. When the mRNA was incubated in the cell-free translational system in the absence of SRP and microsomes, the complete protein with its signal-sequence was produced. The addition of SRP to the incubation mixtures caused protein elongation to cease after 70–100 amino acids had been incorporated. When microsomes containing the SRP receptor also were added to the incubations, the block in protein synthesis was relieved and the complete protein minus the signal sequence was extruded into the lumen of the microsomes.

For further study, see text section "Two Proteins Initiate the Interaction of Signal Sequences with the ER Membrane," p. 697.

8. In yeast, the hydrolysis of ATP by Hsc70 powers the cotranslational import of proteins into the ER lumen. In mammalian cells, ATP hydrolysis is not required for cotranslational translocation. Rather, hydrolysis of GTP during protein synthesis itself is thought to drive the nascent chain across the ER membrane.

For further study, see text section "GTP Hydrolysis Powers Protein Transport into the ER in Mammalian Cells," p. 700.

9. Some integral-membrane proteins (e.g., the asialoglycoprotein receptor) contain an internal hydrophobic sequence of ≈22 residues, which both

directs the nascent protein to the ER membrane and embeds/orients the protein in the membrane. Unlike the N-terminal signal sequences in secretory proteins and other membrane proteins, these dual-function sequences, called internal, signal-anchor sequences, are not cleaved from the nascent protein.

For further study, see text section "A Single Internal Topogenic Sequence Directs Insertion of Some Single-Pass Transmembrane Proteins," p. 704.

10. The protein is initially synthesized as a transmembrane protein with a single stop-transfer membrane-anchor sequence as the transmembrane domain. A short sequence of amino acids in the exoplasmic (luminal) domain, adjacent to the membrane-spanning domain, is recognized by an endoprotease that simultaneously cleaves off the original stop-transfer membrane-anchor sequence and transfers the remainder of the protein to a preformed GPI (glycosylphosphatidlylinositol) anchor in the membrane.

For further study, see text section "After Insertion in the ER Membrane, Some Proteins Are Transferred to a GPI Anchor," p. 705.

11. Improperly folded proteins in the ER have two common fates. One is to aggregate within the ER and form an almost crystalline aggregate. This is the fate of a misfolded mutant form of α_1-antiprotease. The other fate is to be translocated into the cytosol via the translocon and be degraded in the cytosol by the proteosome.

For further study, see text sections "Only Properly Folded Proteins Are Transported from the Rough ER to the Golgi Complex" and "Many Unassembled or Misfolded Proteins in the ER Are Transported to the Cytosol and Degraded," pp. 710–711.

12. Protein disulfide isomerase (PDI) that escapes from the ER is normally retrieved back to the ER from the cis-Golgi reticulum (network) by the KDEL-receptor. PDI contains a C-terminal KDEL sequence. Failure to be retrieved will result in the secretion of escaped PDI from the cell. This will lower the ER accumulation of PDI. KDEL-less PDI will not accumulate elsewhere along the secretory pathway as it lacks targeting information for accumulation in a downstream secretory organelle; rather it is lost to the extracellular medium.

For further study, see text section "ER-Resident Proteins Often Are Retrieved from the Cis-Golgi," p. 711.

13. N-linked oligosaccharide addition is to the amide nitrogen of asparagine. O-linked oligosaccharide addition is to the hydroxyl group of either serine or threonine.

For further study, see text section "Different Structures Characterize N- and O-linked Oligosaccharides," p. 712.

14. Many secreted human proteins are N-glycosylated. Most mammals with the exception of humans and other Old World primates occasionally add a galactose residue, instead of an N-acetylneuraminic acid residue, to galactose, forming a terminal Gal($\alpha1 \rightarrow 3$)Gal disaccharide on some branches of an N-linked oligosaccharide. Because of persistent infections by microorganisms that contain Gal($\alpha1 \rightarrow 3$)Gal disaccharides, human blood always contains antibodies to this disaccharide epitope. Thus Old World primate cells are preferred because of N-linked oligosaccharides produced. Bacteria are a poor choice because they fail to N-glycosylate proteins.

For further study, see text section "Modifications to N-Linked Oligosaccharides Are Completed in the Golgi Complex," p. 718.

15. I–cell disease fibroblasts have M6P receptors. They efficiently internalize lysosomal enzymes containing M6P. I–cell patients lack the GlcNAc phosphotransferase that is required for formation of M6P residues on lysosomal enzymes in the cis-Golgi.

For further study, see text section "Lysomal Storage Diseases Provided Clues to Sorting of Lysosomal Enzymes," p. 720.

16. Regulated secretion occurs only in response to a signal. The proteins to be secreted are stored in special secretory vesicles. Sorting into the regu-

lated secretory pathway is controlled by selective protein aggregation. Constitutive secretion appears to occur by default with secretory proteins, which do not selectively aggregate being included in transport vesicles.

For further study, see text section "Different Vesicles Are Used for Continuous and Regulated Protein Secretion," p. 723.

17. Membrane proteins are sorted to either the apical or basolateral domains by several different mechanisms. One mechanism for targeting proteins to the appropriate domain of the plasma membrane involves sorting in the *trans*-Golgi network. Except for the GPI anchor, which acts as an apical or basolateral targeting signal, no unique sequences have been identified that target proteins to the apical or basolateral domain. Another mechanism operates in hepatocytes where all newly made apical and basolateral proteins are first delivered together from the *trans*-Golgi network to the basolateral membrane. From there, both apical and basolateral proteins are endocytosed to the same vesicles. Within endosomes, there is then sorting and transport to the appropriate domain. The attachment of integral membrane proteins to the cytoskeleton serves as a retention signal and may assist in the apical-basolateral sorting of some proteins. Hence, depending on cell type, at least three different mechanisms for apical-basolateral sorting are possible.

For further study, see text section "Some Proteins Are Sorted from the Golgi Complex to the Apical or Basolateral Plasma Membrane," p. 724.

18. Three different signals that trigger uptake of a protein by receptor-mediated endocytosis have been, at least partially, characterized. These are the sequence Tyr-X-X-ϕ (YXXϕ), Leu-Leu (LL), and an ubiquination signal on some proteins. In all cases, the molecular feature is present in the cytosolic domain of the protein. X = any amino acid, ϕ = bulky hydrophobic residue.

For further study, see text section "Cytosolic Sequences in Some Cell-Surface Receptors Target Them for Endocytosis," p. 728.

19. At the low pH of the late endosomes, the LDL receptor dissociates from its ligand, low-density lipoprotein (LDL), and is recycled to the cell surface. In this acidic compartment, bound iron is released from ferrotransferrin, but the iron-free transferrin, called *apoferritin*, remains associated with the transferrin receptor until the complex is cycled back to the cell surface. Apoferritin is released from the receptor at the neutral pH found there.

For further study, see text section "The Acidic pH of Late Endosomes Causes Most Receptors and Ligands to Dissociate," p. 729.

20. Three classes of coated vesicles have been identified. These are:

COPII mediating anterograde transport from the ER and requiring Sar1 as the small GTPase involved in coat-protein recruitment.

COPI mediating, retrograde transport between the *cis*-Golgi and ER and within the Golgi complex. The small GTPase required is ARF.

Clathrin mediating anterograde transport from the *trans*-Golgi network and endocytosis from the cell-surface. The small GTPase is again ARF.

For further study, see text section "At Least Three Types of Coated Vesicles Transport Proteins from Organelle to Organelle," p. 733.

21. SNARE proteins fall into two classes, V-SNAREs (vesicle SNAREs) and T-SNAREs (target SNAREs). By selective V- and T-SNARE pairing, they are involved in determining the specificity of vesicle fusion with target. In a test tube system using lipsosomes, V- and T-SNAREs alone are sufficient to produce fusion, albeit slowly. Vesicle fusion with target membrane is much more rapid in living cells.

For further study, see text section "Specific Fusion of Intracellular Vesicles Involves a Conserved Set of Fusion Proteins," p. 741.

22a. Comparison of lanes 1 and 2 indicates that pre-LHCP is processed normally in the *in vitro* system. This is a critical part of the experiment because it is imperative to show that this *in vitro* system mimics the *in vivo* situation.

22b. These treatments were used to determine if the mature LHCP is in fact in the thylakoid membrane, where it is found *in vivo,* or whether it is soluble in the stroma or only partially-integrated in the thylakoid membranes. Integral-membrane proteins are resistant to extraction with NaOH and to proteolytic digestion, whereas soluble proteins or partially integrated ones would be susceptible to these treatments.

22c. The protease-treated samples (lanes 4 and 5) migrate slightly faster, indicating a decrease in molecular weight, than the untreated sample (lane 2). This decrease in molecular weight suggests that part of LHCP is exposed to the stroma and thus is susceptible to partial protease degradation.

22d. The data in Figure 19-6 indicate that both ATP and stroma are necessary for binding of preLHCP to isolated thylakoid membranes; this binding protects the bound protein from degradation by protease (lane 2). However, neither ATP or stroma alone can support binding (lanes 3 and 5); in the presence of either one alone, the unbound preLHCP is susceptible to protease degradation (lanes 4 and 6).

23. Protein translocation will occur only when the microsomes are added to the reaction mixture while protein synthesis is actively occurring. An N-terminal signal sequence acts co-translationally. It interacts with SRP while translation is occurring and is threaded through the translocon co-translationally. If the N-terminal signal sequence does not interact rapidly with the membranes, it becomes part of a misfolded protein complex.

24a. Because proteinase K cannot penetrate the microsomal membrane, it can digest only proteins that are not translocated into the lumenal space of the ER. Thus the diffuse bands in Figure 16-3 represent the degradation products of newly synthesized protein that was not translocated during the incubations. The discrete bands represent newly-synthesized protein that was translocated into the ER lumen during the incubations; because this protein was enclosed by a membrane, it was protected from the action of the protease.

24b. Comparison of lanes 1 and 6 indicates that either ATP or GTP or both are necessary for translocation. However, since translocation did not occur in the presence of ATP alone (lane 7), GTP must be a necessary factor. Comparison of lanes 3 and

4 indicates that hydrolysis of ATP but not of GTP is necessary for translocation, since AMP-PNP in the presence of GTP prevented translocation, whereas GMP-PNP in the presence of ATP did s not. Thus although both ATP and GTP are required for translocation, only ATP is hydrolyzed.

Lane 2 indicates that EDTA inhibits translocation of the nascent protein, suggesting that it uncouples the SRP-ribosome complex or prevents ATP hydrolysis, perhaps by chelating Mg^{2+} ions.

Finally, comparison of lanes 1 and 5 suggests that a signal peptidase is present in the translational system. When the incubation time was extended, the peptidase cleaved the 45-kDa nascent protein (lane 1) into the 40-kDa mature protein and a small-signal peptide (lane 5).

25. The results are shown diagrammatically in Figure 17-8.

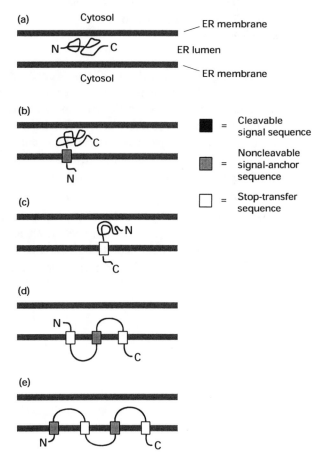

Figure 17-8

26a. All three mutations result in processing (i.e., proteolytic cleavage) of the diK-fusion protein, as evidenced by the presence of bands representing lower-molecular-weight cleavage products. There are some variations in the extent of processing among the mutants, but these are relatively minor. These results suggest that all three mutations prevent retrieval of the diK-fusion protein.

26b. The three mutations have divergent effects on secretion of CPY. The *ret1-1* and *sec21-2* mutations have little effect, as evidenced by the strong band corresponding to mCPY at both the permissive and nonpermissive temperatures. In contrast, the *sec21-1* mutation inhibits CPY secretion, as indicated by the weak mCPY band at the nonpermissive temperature. Of these three mutations, *ret1-1* and *sec21-2* were in fact selected on the basis of their effect on protein retrieval, whereas *sec21-1* was selected on the basis of its effect on protein secretion.

26c. The finding that two different alleles of *sec21* differ in phenotype suggests that the *sec21* protein must have at least two different functional domains. Since only one allele of *ret1* has been isolated so far, nothing can be concluded about the possibility of different domains in the Ret1 protein based on the available genetic evidence.

26d. These mutations suggest that COP proteins and the coatomer play a role in both protein retrieval to the ER and protein secretion from the ER. Whether the role in both is direct or indirect is open to question.

27a. Since only VSV G protein labeled with [³H]palmitate is detected in this experiment, all the G protein revealed on the electrophoretograms initially had resided in donor *cis*-Golgi membranes. This G protein is susceptible to endoglycosidase H. The acquisition of resistance to this enzyme during the *in vitro* incubation is indicated by the appearance of the resistant G_R band. Since this resistance depends on processing in the acceptor *medial*-Golgi membranes, appearance of the G_R band indicates that the labeled G protein is part of a structure which had transformed from the *cis*- to *medial*-Golgi membrane properties.

27b. As the *in vivo* chase time increases, there is a profound decrease in the proportion of prelabeled G protein that is processed (i.e., acquires endoglycosidase H resistance). This finding suggests that as the chase time is increased, labeled, incompletely processed G protein is found in a subcompartment that is incompetent in the *in vitro* transport assay.

27c. The results suggest that once G protein has moved from one Golgi subcompartment to another *in vivo*, it cannot transfer back during the *in vitro* transport assay for the completion of a given enzymatic processing step on its sugar-side chains. In other words, transport through the Golgi is unidirectional. Of course, this conclusion assumes that most of the labeled G protein is present within the Golgi and does not transfer out of the Golgi during the experiment. Experiments to verify this assumption (e.g., by immunolabeling) could be done.

18 Cell Motility and Shape I: Microfilaments

PART A: *Chapter Summary*

Actin filaments (also known as microfilaments or F-actin) are dynamic polymers assembled from globular actin subunits (G-actin). Actin filaments are involved in generation and alteration of cell shape and in cell locomotion. Muscle is the best understood example of cell motility, and study of muscle cells has provided much of our knowledge of actin and actin-binding proteins such as myosin. However, non-muscle cells also contain significant amounts of actin and actin-binding proteins, and forces produced by actin polymerization as well as by myosin motors are important for many processes in nonmuscle cells.

PART B: *Reviewing Concepts*

18.1 *The Actin Cytoskeleton*
(lecture date _____)

1. There are four states of actin. What are the four states and what is the significance of each?

2. What structural feature of actin filaments can be visualized by use of filaments "decorated" with myosin S1 fragments?

3. What is the major structural difference between the actin-binding proteins found in actin bundles and networks? How does this difference contribute to the characteristic structures of bundles and networks?

4. To generate and maintain cell shape, actin filaments must interact with the plasma membrane. How are actin filaments connected to the plasma membrane of platelets, muscle cells, and epithelial cells?

18.2 *The Dynamics of Actin Assembly*
(lecture date _____)

5. Actin polymerization is accompanied by ATP hydrolysis, but it is thought that ATP hydrolysis is not required for polymerization. What is the experimental basis for this conclusion?

6. What do plasma membrane extension, intracellular bacteria and virus infection, and the sperm acrosome reaction all have in common?

7. What does gelsolin do to actin filaments? How does the cell control gelsolin activity and for what purpose?

8. Cells contain much more G-actin than expected from in vitro assembly experiments. What is thought to account for the high cellular concentration of G-actin?

18.3 *Myosin: The Actin Motor Protein*
(lecture date _____)

9. How is myosin thought to move along actin filaments?

10. There are at least thirteen different types of myosin motors. What do they have in common and what makes each type distinct?

11. What molecular feature allows two myosin heavy chains to form a dimer?

12. The endoplasmic reticulum (ER) moves in one direction along bundles of actin filaments in the green

alga *Nitella*. What is the most probable reason why movement does not occur in both directions?

18.4 *Muscle, A Specialized Contractile Machine* (lecture date _____)

13. Microinjection of an antibody to myosin light-chain kinase inhibits the contraction of vertebrate smooth muscle but not that of vertebrate skeletal muscle. Contraction of both types of muscle, however, is associated with a rise in cytosolic Ca^{2+}. Explain these findings.

14. Diagram a sarcomere, including actin filaments, myosin filaments, titin filaments, nebulin filaments, and Z disks. Where would α-actinin, CapZ, and tropomyosin be located?

15. What is the role of capping proteins in the sarcomere? What is the role of titin and nebulin in the sarcomere?

18.5 *Actin and Myosin in Nonmuscle Cells* (lecture date _____)

16. Contractile bundles occur in nonmuscle cells although these structures are less organized than the sarcomeres of muscle cells. What is the purpose of nonmuscle contractile bundles?

17. What is the experimental evidence for the involvement of myosin II in cytokinesis?

18. Myosins play important roles in motile processes other than contraction. What other functions do myosins provide, and what types of myosins are involved?

18.6 *Cell Locomotion* (lecture date _____)

19. Keratinocyte movement has been extensively characterized as a model of cell locomotion. What steps are thought to be involved in this movement?

20. What roles do Ras-related G proteins and Ca^{2+} play in cell locomotion?

PART C: *Analyzing Experiments*

21. (•) Understanding of actin filaments has been greatly facilitated by the ability of scientists to purify actin and actin-binding proteins and the ability to assemble actin filaments in vitro. Below are various experimental approaches designed to characterize actin assembly and the effects of actin-binding proteins on this assembly.

a. The graph in Figure 18-1a depicts the actin polymerization rate at the plus (+) and minus (−) ends of rabbit actin as a function of actin concentration. Assume that you could add microfilaments of a predefined length to rabbit actin maintained at the concentrations labeled A, B, and C in this figure. Diagram the appearance of the filaments after a 10-min incubation at each of the indicated actin concentrations, if the original filaments are depicted as follows:

Original filament: + _____ −

Make sure to mark the location of the original (+) and (−) ends of the filament on your diagrams.

b. A novel actin-binding protein (X) is overexpressed in certain highly malignant cancers. You wish to determine if protein X caps actin filaments at the (+) or (−) end. You incubate an excess of protein X with various concentrations of G-actin under conditions that induce polymerization. Control samples are incubated in the absence of protein X. The results are shown in Figure 18-1b. How can you conclude from these data that protein X binds to the (+) end of actin filaments? Design an experiment, using myosin S1 fragments and electron microscopy, to corroborate the conclusion that protein X binds to the (+) end. What results would you expect if this conclusion is correct?

c. An in vitro system was developed to study actin assembly and disassembly in nonmuscle cells. In this study, tissue culture cells were incubated for several hours with [^{35}S]methionine so that all the actin monomers in each filament were labeled. Actin filaments were then collected by differential centrifugation and put into a buffer containing one of three different cytosolic extracts (A, B, or C). The amounts of soluble actin in each sample was monitored over time (see Figure 18-1c). What do these data indicate about the effects of A, B, and C on the assembly and disassembly of actin filaments?

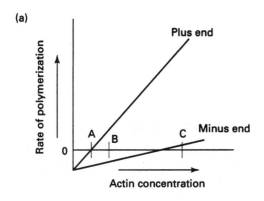

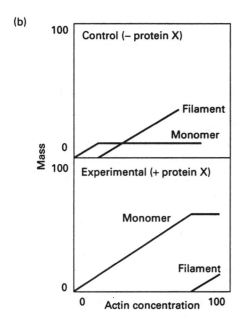

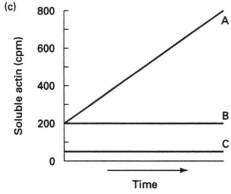

Figure 18-1

tuted with alanines. Mutant 1 had two substitutions and Mutant 2 had three substitutions. To test the effect of wild type cofilin and the mutant forms of cofilin on actin filaments in vivo, the genes for cofilin and the two cofilin mutants were introduced independently into yeast cells. Yeast cells contain "actin patches" where F-actin structures associated with the cell cortex. The cells containing the introduced genes were then treated with a drug that sequestered G-actin and the effect on actin patches was determined (Figure 18-2a).

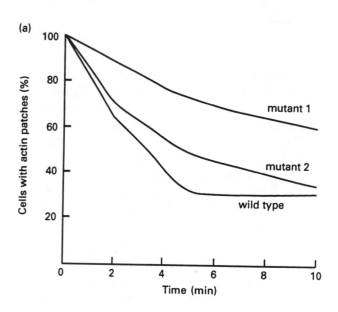

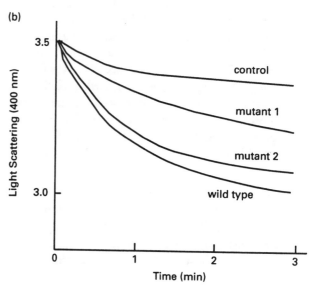

Figure 18-2

22. (••) The actin-binding protein cofilin is thought to be involved in the regulation of actin dynamics in cells. To investigate the action of cofilin, in vivo and in vitro experiments were conducted with wild-type cofilin and two cofilin mutants. In both mutants, a different cluster of charged amino acids was substi-

a. What was the purpose of treating the cells with a G-actin sequestering drug?

b. What effect did wild type cofilin and the two mutant proteins have on actin patches? What can you infer about the action of cofilin?

c. To test the effect of wild type cofilin and the mutant forms of cofilin on actin filaments in vitro, the proteins were expressed, purified, and each was combined with actin filaments. The effect of each protein on the actin filaments was then monitored by light scattered (Figure 18-2b). What do you think the light scattering measures? How do the in vitro data correlate with the in vivo data?

23. (•••) Although many of the effects of actin are mediated at the plasma membrane, most known actin-binding proteins are soluble. An exception is ponticulin, a 17-kD membrane glycoprotein from *Dictyostelium discoideum,* which has been shown to traverse the membrane of this cellular slime mold. Membrane preparations from *Dictyostelium* bind F-actin and nucleate actin polymerization at the membrane surface. This process appears to involve the generation of actin trimers at the cytoplasmic surface; these trimers have both the pointed and barbed ends free to elongate.

Ponticulin can be purified by extraction of ameboid cells of *Dictyostelium* with the nonionic detergent Triton X-114 to generate a purified cytoskeletal preparation, and this preparation was then extracted with another detergent, octylglucoside, which removed the ponticulin from the cytoskeletons. F-actin affinity chromatography, high-pressure liquid chromatography (HPLC), and hydrophobic interaction chromatography (HIC), all in the presence of octylglucoside, were used to purify the ponticulin further. SDS-PAGE analysis of ^{125}I-labeled preparations indicated that these preparations contained only very minor amounts of contaminating proteins, and no proteins that comigrated with actin.

a. An important question was whether this purified detergent-soluble preparation retained actin-binding activity. To answer this question, combinations of F-actin and ^{125}I-labeled ponticulin were incubated for 1 h at room temperature, and these reaction mixtures were then centrifuged to pellet the actin filaments. Pellets (P) and supernatants (S) were analyzed by SDS-PAGE and autoradiography. Results are shown in Figure 18-3a. Based on these results, can you conclude that this ponticulin preparation retains actin-binding activity?

b. Membrane proteins solubilized in octylglucoside can be incorporated into membrane vesicles after dilution of the detergent in the presence of excess lipid; this process is called reconstitution. Purified detergent-soluble ponticulin was reconstituted in membranes containing *Dictyostelium* lipids or a synthetic phosphatidylcholine (DMPC); the vesicles then were assayed for the ability to induce polymerization of G-actin. Controls included determination of G-actin polymerization in vesicles reconstituted in the absence of lipid and in lipid-containing vesicles of both types prepared without ponticulin. The results of these assays are shown in Figure 18-3b. What do these data allow you to conclude about the ability of various membrane lipids to support ponticulin-induced actin polymerization? What experimental artifacts might account for the observed lipid effects?

c. Another possible explanation for the lipid specificity observed in Figure 18-3b is that *Dictyostelium* lipids, but not DMPC or other lipids, promote G-actin polymerization via generation of ponticulin clusters, which provide multivalent actin-binding sites. To test this hypothesis, the nucleation activity of ponticulin reconstituted with *Dictyostelium* lipids at a wide range of lipid:protein ratios was assayed. Polymerization activity was measured as in part (b), yielding data similar to that shown in Figure 18-3b. From these data the polymerization rates at different lipid:protein ratios were calculated and then divided by the rate observed for actin alone, yielding fold increases as shown in Figure 18-3c. Final concentrations of *Dictyostelium* lipids were 1.4 μM, 14 μM, or 70 μM in the assay mixture. Do the data in Figure 18-3c support the hypothesis that membrane-bound ponticulin acts as an oligomer in nucleation of G-actin polymerization? Explain your answer.

(a)

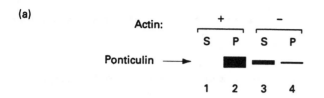

(b)

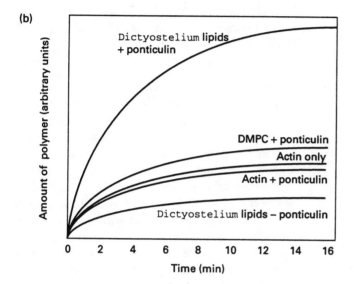

(c)

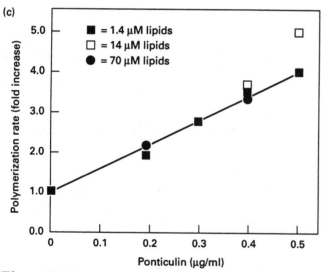

Figure 18-3

Part D:*Answers*

1. The four states of actin are ATP-F-Actin, ATP-G-Actin, ADP-F-Actin, and ADP-G-Actin. The F-actin designation indicates that the actin subunit is part of a filament, while the G-actin designation indicates that the subunit is not part of a filament (i.e., it is unassembled or free in solution/cytosol). The ATP and ADP designations indicate the nucleotide bound to the actin subunit.

For further study, see text section "ATP Holds Together the Two Lobes of the Actin Monomer," p. 753.

2. The observation that all bound S1 fragments point in one direction [i.e., toward the minus (–) end] demonstrates the polarity of actin filaments and provides a method for distinguishing the two ends.

For further study, see text section "F-Actin Has Structural and Functional Polarity," p. 754.

3. All the actin-binding proteins in bundles and networks have two actin-binding sites and thus they can cross-link a pair of actin filaments. Cross-linking proteins that have a short inflexible domain between the actin-binding sites hold the filaments closely together in a nearly parallel alignment, forming a bundle. Other cross-linking proteins have a long, flexible domain between the actin-binding sites. These proteins tend to hold actin filaments farther apart, and in orthogonal arrays, forming a network.

For further study, see text section "The Actin Cytoskeleton is Organized into Bundles and Networks of Filaments," p. 755.

4. In these cells, the actin filaments of the cell cortex are indirectly connected to transmembrane proteins by actin-binding proteins. In platelets, filamin serves to connect actin filaments to the glycoprotein 1b-IX complex (which also interacts with extracellular blood clotting proteins). In muscle, dystrophin connects actin filaments to a glycoprotein complex (which also interacts with proteins in the extracellular matrix). In epithelial cells, ezrin and the adapter protein EBP50 connect actin filaments to the CFTR and other cell surface receptors.

For further study, see text section "Cortical Actin Networks Are Connected to the Membrane," p. 756, and Figure 18-9, p. 759.

5. G-actin containing ADP or AMPPNP (a nonhydrolyzable ATP analog) is able to form filaments, so hydrolysis of ATP is not essential for polymerization of actin subunits.

   For further study, see text section "Actin Polymerization in Vitro Proceeds in Three Steps," p. 761.

6. All involve movement driven by actin polymerization.

   For further study, see text section "Many Movements Are Driven by Actin Polymerization," p. 766.

7. Gelsolin severs actin filaments into shorter fragments, and caps the (+) ends of the resulting fragments thereby preventing addition of new subunits. Gelsolin activity is sensitive to Ca^{2+} and cells can regulate gelsolin's function by regulating Ca^{2+} levels in localized regions of the cell's cortex. Gelsolin may play a role in gel-sol transitions during ameboid movement.

   For further study, see text section "Some Proteins Control the Lengths of Actin Filaments by Severing Them," p. 765.

8. Actin-sequestering proteins such as thymosin β_4 bind to G-actin and hold it in a form that is unable to polymerize.

   For further study, see text section "Inhibition of Actin Assembly by Thymosin β_4," p. 764.

9. Myosin uses the energy of ATP hydrolysis to walk along an actin filament in discrete steps of 5-10 nm. Currently, it is not clear if myosin takes one step per ATP hydrolyzed but this is assumed in most models of movement. The head domain is the region of the motor that interacts with actin and that binds and hydrolyzes ATP. The head domain is thought to go through a cycle of repeating conformational changes that depend on the nucleotide state of the myosin and myosin binding to actin. These conformational changes in the head are transmitted and amplified to the rest of the molecule as part of the power stroke that generates movement.

   For further study, see text section "Myosin Heads Walk along Actin Filaments," p. 770; "Myosin Heads Move in Discrete Steps, Each Coupled to Hydrolysis of One ATP," p. 771; and "Conformational Changes in the Myosin Head Couple ATP Hydrolysis to Movement," p. 773.

10. All myosins use energy derived from ATP hydrolysis to "walk" along actin filaments. Depending on the specific type of myosin, this movement is used to generate contraction or to transport specific cellular components relative to actin filaments. All myosins are composed of one or two heavy chains (the motor subunit) and several light chains. The heavy chains of all types of myosin have similar head domains (which interact with actin and generate force to move) but have different tail domains which provide specific functions to each type of myosin. In addition, the different types of myosins may have different light chains that act to regulate motor activity.

    For further study, see text section "All Myosins Have Head, Neck, and Tail Domains with Distinct Functions," p. 769.

11. The tail domain of dimeric myosin motors is α helical, and the α helices of two heavy chains associate to form a rodlike coiled-coil structure.

    For further study, see text section "All Myosins Have Head, Neck, and Tail Domains with Distinct Functions," p. 769.

12. Movement of the ER in *Nitella* results from myosin-actin interactions, whose orientation is dictated by the polarity of the actin filaments. Because all the actin filaments run in one direction, the cytoplasm is propelled in one direction only.

    For further study, see text section "Myosin Heads Walk along Actin Filaments," p. 770, and Figure 18-40, p. 787.

13. The mechanism by which a rise in Ca^{2+} triggers contractions differs in skeletal and smooth muscle. In skeletal muscle, binding of Ca^{2+} to troponin C leads to muscle contraction. In comparison, contraction of smooth muscle is triggered by activation of myosin light-chain kinase by Ca^{2+}-calmodulin. Thus only smooth muscle is inhibited by microinjection of antibodies to myosin light-chain kinase.

For further study, see text section "Role of Tropomyosin and Troponin in Skeletal Muscle Contraction," p. 780; and "Activation of Myosin by Calcium-Dependent Phosphorylation," p. 781.

14. See Figure 18-4. The proteins α-actinin and CapZ would be found at/near the Z disks, while tropomyosin would be found along the length of the actin thin filaments.

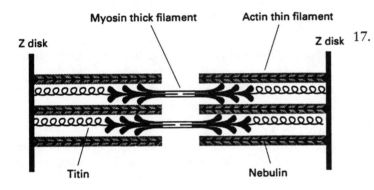

Figure 18-4

For further study, see text section "Skeletal Muscles Contain a Regular Array of Actin and Myosin," p. 775.

15. Capping proteins function to keep actin thin filaments length constant in the sarcomere. Tropomodulin performs this function at the (−) ends near the sarcomere center, and CapZ stabilizes the (+) ends at/near the Z disks. Titin is an extremely large, elastic protein that acts to keep myosin filaments centered between the Z disks during contraction or stretching. Nebulin forms long nonelastic filaments that organize the actin thin filaments and may also act as a molecular ruler to regulate actin filament length in the sarcomere.

For further study, see text section "Thin Filaments," p. 776, and "Titin and Nebulin Filaments Organize the Sarcomere," p. 778.

16. The principal contractile bundles of nonmuscle cells are the circumferential belt of epithelial cells, the stress fibers present on cells cultured on artificial substrates, and the contractile ring. The circumferential belt allows a cell to control its shape and a sheet of epithelial cells to move as a unit (as it would in healing or during development). Stress fibers appear to function in cell adhesion rather than the movements associated with cell locomotion. The contractile ring generates the cleavage furrow during cytokinesis that eventually leads to division of a single cytoplasm into two.

For further study, see text section "Actin and Myosin II Are Arranged in Contractile Bundles That Function in Cell Adhesion," p. 783.

17. When the myosin II gene was deleted, myosin II expression was inhibited by antisense RNA techniques, or the myosin II protein was inhibited by function-blocking antibodies, cells failed to assemble a contractile ring and became multinucleate (because chromosome segregation but not cytokinesis occurred).

For further study, see text section "Actin and Myosin II Have Essential Roles in Cytokinesis," p. 784.

18. In addition to contraction, certain myosins are involved in other motile processes including vesicle transport and membrane movement at the leading edge. Myosin I and myosin V have been shown to move vesicles along actin filaments and myosin I has been implicated in movements at the leading edge.

For further study, see text section "Membrane-Bound Myosins Power Movement of Some Vesicles," p. 785.

19. The movement of keratinocytes is pictured to consist of four steps. First, the leading edge of the cells is extended (perhaps by actin polymeriza-

tion). Second, the newly extended membrane forms an attachment to the substrate, which anchors this portion of the cell to the substrate and prevents its retraction. Third, the bulk of the cell body is translocated forward by an unknown mechanism. Finally, the focal adhesion at the rear of the cell are broken so that the tail-end of the cell is brought forward.

For further study, see text section "Controlled Polymerization and Rearrangements of Actin Filaments Occur during Keratinocyte Movement," p. 787.

20. Ras-related G proteins are involved in the signal-transduction pathways that are activated in fibroblasts as part of the wound healing response. Depending on the specific G protein, activation of this type of signal pathway may lead to formation of filopodia, lamellipodia, focal adhesions, and/or stress fibers. Ca^{2+} is probably involved in activation of gelsolin and contraction of myosin II, and may be found in intracellular gradients important for steering in chemotactic cells.

For further study, see text section "Migration of Cells Is Coordinated by Various Second Messengers and Signal-Transduction Pathways," p. 790.

21a. At concentration A, there would be no growth at the (+) end and shortening at the (−) end. At concentration B, there would be growth at the (+) end and shortening at the (−) end but less than at concentration A. At concentration C, there would be growth at both ends, but more at the (+) end.

21b. In the presence of protein X, the concentration of actin needed to promote polymerization is shifted to a much higher concentration. This is the expected finding if protein X binds to (+) end of microfilaments. If the protein bound to the (−) end, one would observe a shift of much less magnitude. Myosin-decorated microfilaments, in the presence or absence of protein X, could be incubated with high concentrations of actin. If protein X specifically binds to and inhibits addition at the (+) end, filament growth should be visible at the nonbarbed (−) end, but not at the barbed (+) end.

21c. Extract A causes a net depolymerization of actin filaments, as evidenced by the steady increase in the amount of soluble labeled actin over time. In the presence of extract B, the actin filaments maintain a constant length; however, they must be losing and adding actin filaments at identical rates, otherwise there would be no soluble actin monomers in solution. The filaments in extract C are likewise maintaining a constant length; the low basal level of labeled monomeric actin in this case suggests that C contains an unidentified capping entity that binds to the actin filaments and inhibits their assembly and disassembly.

For further study, see text section "The Dynamics of Actin Assembly," p. 761.

22a. The G-actin sequestering drug was used to eliminate G-actin addition to the cell's actin filaments, and thus permitted examination of cofilin effects on actin disassembly.

22b. Cofilin and the mutant cofilin proteins (to lesser extents) promoted the loss of actin patches. The results suggest that cofilin promotes the disassembly of actin filaments.

22c. The light is scattered by actin filaments and serves as a measure of the amount of actin filaments present. As the filaments disassemble as shown in the figure, the amount of light scattered decreases. The in vitro data correlate well with the in vivo data—both suggest that cofilin promotes actin disassembly.

For further study, see text section "Some Proteins Control the Lengths of Actin Filaments by Severing Them," p. 765.

23a. In the presence of F-actin, the ^{125}I-labeled ponticulin was found in the pellet (lane 2), whereas very little was found in the supernatant (lane 1). In the absence of actin, most of the ponticulin was found in the supernatant (lane 3), but a small amount was found in the pellet (lane 4). This material probably is aggregated ponticulin; such behavior is not unusual for intrinsic membrane proteins even in the presence of detergent. These results indicate that purified soluble ponticulin retained actin-binding activity, although the aggregation behavior makes a precise determination of the dissociation constant somewhat problematic.

23b. Ponticulin-mediated actin-nucleation activity is apparently very dependent on the composition of the lipid bilayer. The rate of actin polymerization in DMPC + ponticulin bilayers was barely elevated over that of actin alone, whereas vesicles containing *Dictyostelium* lipids + ponticulin stimulated polymerization substantially. Possible artifacts that might explain this lipid specificity include lesser incorporation of ponticulin in the DMPC membranes, different curvature of the different membrane vesicles, or differences in physical state (fluid or solid or hexagonal phase) of the vesicles. Control experiments and experiments with a variety of other lipids appeared to rule out all of these artifactual explanations.

23c. These data do not support the hypothesis. In this assay, the rate of G-actin polymerization would be expected to be linearly dependent upon protein concentration for a monomer, and to be proportional to the square of the protein concentration for a dimer in equilibrium with a monomer. Higher-order dependence would be observed for functional oligomers greater than $n = 2$. The observed linear dependence of the polymerization rate on protein concentration indicates that membrane-bound ponticulin acts as a monomer, not an oligomer, in actin nucleation. This conclusion is also consistent with the observation (data not shown) that ponticulin in *Dictyostelium* lipid vesicles is resistant to chemical cross-linking. Other possible explanations for the lipid specificity indicated in Figure 18-3c are that some particular lipid (or proteolipid) serves as a cofactor for nucleation or provides a specific environment necessary for functional association of ponticulin with actin. These hypotheses have not yet been experimentally tested.

For further study, see text section "The Dynamics of Actin Assembly," p. 761.

19 Cell Motility and Shape II: Microtubules and Intermediate Filaments

PART A: *Chapter Summary*

Microtubules are filaments of tubulin subunits. These dynamic polymers are important in intracellular organization, vesicle transport, organelle positioning, and in chromosome segregation during mitosis. Microtubules also can be found in cellular appendages such as cilia and flagella, where the bending action of microtubule bundles produces the force to power cell swimming. The various functions of microtubules are orchestrated by many different kinds of microtubule-associated proteins.

Intermediate filaments are less dynamic than microtubules and are not involved in cell motility, but rather serve an important role in providing structural support to the plasma membrane and to the nucleus.

PART B: *Reviewing Concepts*

19.1 *Microtubule Structures*
(lecture date _____)

1. Compare α- and β-tubulin in terms of nucleotide binding and hydrolysis.

2. What is the underlying basis for microtubule polarity?

3. What are the main functions of MTOCs? What are the main components of MTOCs?

4. What is the functional reason why some microtubules undergo periodic disassembly and reassembly, whereas others are quite stable and exhibit little cycling between the assembled and disassembled states?

19.2 *Microtubule Dynamics and Associated Proteins* (lecture date _____)

5. What is the effect of temperature on microtubule assembly? How does the presence of short microtubule fragments affect tubulin assembly?

6. MAPs were first discovered as nonspecific "contaminating" proteins in purified tubulin preparations. What is the evidence that led scientists to conclude that these proteins are actually specifically associated with microtubules and tubulin?

7. Naturally produced poisons such as colchicine and taxol are effective inhibitors of mitosis. How do these compounds inhibit cell division and which major disease can such toxins be used to treat?

8. Microtubules both in vitro and in vivo undergo dynamic instability, and this type of assembly is thought to be intrinsic to the microtubule. What is the current model to account for dynamic instability?

19.3 *Kinesin, Dynein, and Intracellular Transport* (lecture date _____)

9. How was kinesin first characterized and isolated?

10. Various studies on axonal transport have demonstrated the following: (a) Kinesin is a (+) end–directed motor protein; (b) all the microtubules in an individual axon have the same polarity; (c) vesicles can move in both directions simultaneously in an individual axon. How can these seemingly discrepant observations be resolved?

11. There are at least 12 different members of the kinesin family of microtubule motor proteins. What structural feature makes each member distinct and what is the functional consequence of this difference?

19.4 *Cilia and Flagella: Structure and Movement* (lecture date _____)

12. Describe the experimental results demonstrating that the outer doublets in axonemes slide during motion of cilia and flagella.

13. Individuals who are genetically defective in dynein demonstrate a characteristic cough, often have chronic bronchitis, and are sterile. What is the molecular basis for these clinical symptoms?

14. What structure serves as a template for axoneme assembly?

15. Dynein generates sliding movement but axonemes and therefore cilia and flagella undergo bending movements. How is sliding movement converted into bending movement?

19.5 *Microtubule Dynamics and Motor Proteins during Mitosis* (lecture date _____)

16. The mitotic spindle is a bipolar structure. How is this structure generated as a cell prepares to divide?

17. The poleward movement of kinetochores (and hence chromatids) during anaphase A requires that the kinetochore maintain a hold on the shortening microtubule. How does a kinetochore hold onto shortening microtubules?

18. What models have been proposed to account for the separation of spindle poles during anaphase B?

19.6 *Intermediate Filaments* (lecture date _____)

19. How is the structure of intermediate filament subunits related to the function of the assembled filaments?

20. How has our increased knowledge about the different protein classes that compose intermediate filaments (IFs) contributed to progress in cancer diagnosis and treatment?

21. Although not considered as dynamic as actin filaments or microtubules, intermediate filament assembly and disassembly is routinely regulated by cells. What mechanism is used to control assembly of intermediate filaments?

PART C: *Analyzing Experiments*

19.3 *Kinesin, Dynein, and Intracellular Transport*

22. (••) The microtubule motor kinesin exhibits processive movement, meaning that a single motor can move for many micrometers along a microtubule without falling off. This property may be important in vivo in cases where a small vesicle with only a single kinesin bound must travel long distances on the scale of the cell. One model to account for the processive nature of kinesin is termed the hand-over-hand model. A key feature of this model is that one head should always be attached to the microtubule, which would lessen the chance of the motor falling off the microtubule.

a. To test this model, a single-headed kinesin was prepared and its motile properties examined. If the hand-over-hand model is correct, would you expect the single-headed kinesin to be processive?

b. Figure 19-1 depicts the landing rate of microtubules on a coverslip coated with either two-headed or one-headed kinesin at different motor densities (landing is defined as interacting with and being moved by motors attached to the glass surface). What can you conclude from this data?

c. Another feature of the hand-over-hand model is that one head, upon binding to the microtubule, promotes the release of the other head from the microtubule. If this hypothesis is correct, what would you expect when a one-headed kinesin molecule bound to a microtubule?

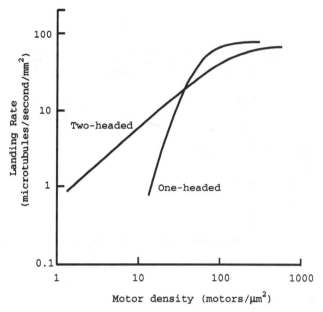

Figure 19-1

23. (••) Motor proteins are generally considered to move cellular components along microtubules. However, one group of kinesin family motors, termed Kin I kinesins, may have other functions as described below.

   a. In one set of experiments, microtubules were observed in a microscope chamber in which the solution could be rapidly exchanged. Figure 19-2a depicts the effect on microtubule length as a function of time when the tubulin and buffer in the chamber was replaced first by an identical tubulin and buffer solution and then by buffer alone. Figure 19-2b depicts the results when the tubulin and buffer solution in the chamber was replaced with a solution containing tubulin and the Kin I motor protein. What conclusions can you draw from these experiments?

   b. In another assay, taxol-stabilized microtubules were incubated alone (Ctrl), with kinesin heavy chain (KHC) and ATP, with the Kin I motor and ATP, or with the Kin I motor and AMPPNP, and then centrifuged to pellet polymerized tubulin. Supernatant (s) and pellet (p) fractions were then separated by SDS-PAGE and the gel was stained to reveal the position of tubulin (Figure 19-2c).

What conclusions can you draw from these experiments? What is the significance of the nucleotide present for Kin I activity? Similar results were obtained when GMPCPP, a nonhydrolyzable GTP analog, was used to assemble stabilized microtubules. What does this result suggest about the action of Kin I motors?

   c. There is additional evidence that Kin I motors can produce similar results even at low ratios relative to tubulin. Given this fact and what you know about microtubule assembly, where might you expect to find the motor on microtubules?

(a)

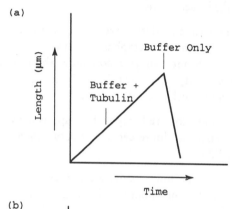

(b)

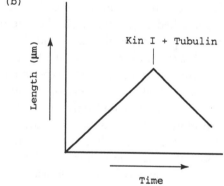

(c)

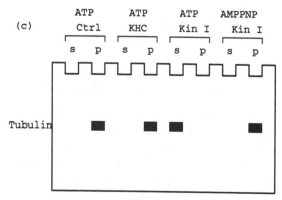

Figure 19-2

19.5 *Microtubule Dynamics and Motor Profiles during Mitosis*

24. (••) Microtubule dynamics in cells is a very active area of research. One approach to characterizing cellular microtubule assembly and disassembly involves the injection into cells of tubulin that has been covalently labeled with a fluorescent dye. This fluorescent tubulin is freely incorporated into microtubules in the injected cells. rendering the microtubules visible in a fluorescence microscope. Using a powerful laser, it is possible to irradiate a small region of the microtubules, bleaching the fluorescent dye so that it no longer is fluorescent and then to measure the recovery of fluorescence in the bleached area.

   a. In one experiment, data were collected for cells in interphase and in metaphase as shown in Figure 19-3a. What can you conclude about the relative stability of microtubules in metaphase and interphase cells from this data?

   b. What mechanisms are likely to be responsible for the differences in fluorescence recovery shown in Figure 19-3a?

   c. In another experiment, cells were microinjected with X-rhodamine tubulin, which labels all microtubules. Areas of the mitotic spindle were then photobleached to determine (1) the rate of movement of the photobleached area and (2) the degree of fluorescence recovery of the photobleached area. One of the first experiments this group performed was to develop a buffer that lysed cells and preserved only kinetochore fiber microtubules. Why was this a critical experiment? What does the effect of this buffer suggest about the differences in microtubules in vivo? What other methods might be useful for preserving kinetochore tubules at the expense of other microtubules?

   d. When anaphase cells were labeled as described in (c) and then photobleached, the expected absence of fluorescence was noted. How could you demonstrate that the irradiation does not destroy the kinetochore microtubules at this point?

   e. When the positions of the photobleached spot was monitored in anaphase cells, it moved 1 μm/100 s toward the pole, but no fluorescence recovery of the photobleached area was noted. What is the interpretation of these findings?

   f. Figure 19-3b shows the fluorescence recovery after photobleaching of anaphase and metaphase kinetochore tubules. What is the significance of these data?

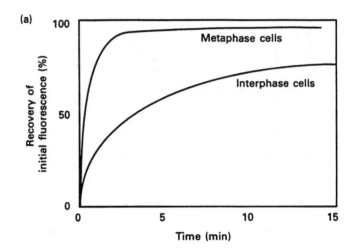

(a)

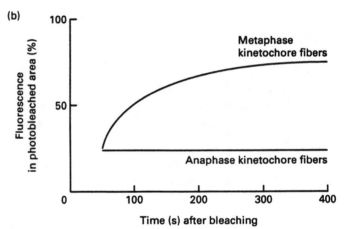

(b)

Figure 19-3

Answers

1. Both α- and β-tubulin bind GTP, and this form of the dimer is competent to assemble into microtubules. However, only β-tubulin is able to hydrolyze the bound GTP to GDP (which occurs after assembly into a microtubule) and exchange GDP for GTP (after the subunit disassembles from a microtubule). The GTP bound to α-tubulin is never hydrolyzed or exchanged with free nucleotide.

For further study, see text section "Heterodimeric Tubulin Subunits Compose the Wall of a Microtubule," p. 796.

2. The microtubule is a polar filament because the tubulin heterodimer is polar and these subunits assemble head-to-tail to form the microtubule wall. Thus one end (−) of the microtubule has α-tubulin exposed while the other end (+) has β-tubulin exposed.

For further study, see text section "Heterodimeric Tubulin Subunits Compose the Wall of a Microtubule," p. 796.

3. MTOCs (or centrosomes) serve to nucleate and organize microtubules. Almost all microtubules originate from the MTOC and the (−) ends of these microtubules remain associated with the MTOC. All MTOCs are composed of pericentriolar material (a lattice of microtubule-associated proteins) and many animal MTOCs contain centrioles. The pericentriolar material is the site of microtubule nucleation, which depends on γ-tubulin ring complexes. It is not clear what function the centrioles serve, but it is not as templates for microtubule assembly.

For further study, see text section "Microtubules Assemble from Organizing Centers," p. 799, and "The γ-Tubulin Ring Complex Nucleates Polymerization of Tubulin Subunits," p. 800.

4. Microtubules must disassemble and reassemble in response to the changing structural requirements of the cell. This is especially true in the case of spindle microtubules, which are assembled and then disassembled during the cell cycle. The microtubules in flagella, in contrast, are quite stable. Cytoplasmic and flagellar microtubules are associated with different MAPs, which may be related to microtubule stability. Additionally, experiments with colchicine and cold-induced depolymerization suggest that the differences in the stability of various microtubules may also be related to the presence of different tubulin isotypes.

For further study, see text section "Microtubules Form a Diverse Array of Both Permanent and Transient Structures," p. 797.

5. Microtubule assembly is temperature dependent; warm temperatures (37°C) favor microtubule assembly, while cold temperatures (4°C) favor microtubule disassembly. Short microtubule fragments accelerate tubulin assembly because the fragments act as nuclei for assembly and would allow the lag phase of assembly to be bypassed.

For further study, see text section "Microtubule Assembly and Disassembly Occurs Preferentially at the (+) End," p. 802.

6. MAPs copurify with tubulin and maintain the same quantitative ratio of MAP:tubulin throughout the purification process. Additional evidence comes from the similar distribution of specific MAPs (stained by specific antibodies) and microtubules.

For further study, see text section "Assembly MAPs Cross-Link Microtubules to One Another and Other Structures," p. 807.

7. Colchicine inhibits cell division by blocking assembly of tubulin into microtubules while taxol inhibits cell division by stabilizing microtubules. Although the effects are opposite, the results are the same: failure of the mitotic spindle to function properly and therefore failure of the cell to divide. These poisons and related compounds have been used to treat cancer, which can be considered uncontrolled cell division.

For further study, see text section "Colchicine and Other Drugs Disrupt Microtubule Dynamics," p. 806.

8. During dynamic instability, microtubules alternate between growth and shortening. The current model for dynamic instability is the GTP cap model. According to this model, GTP-tubulin (subunits with GTP bound to β-tubulin) can add to the end of a growing microtubule, but sometime after assembly, the GTP will be hydrolyzed to GDP, leaving GDP-tubulin, which makes up the bulk of the microtubule. Thus GTP-tubulin is present only at the microtubule end, and as long as this situation holds, the microtubule will continue to grow because the cap stabilizes the entire

microtubule. However, if GDP-tubulin becomes exposed at the end then the stabilizing cap is lost and the microtubule will begin to shorten. The microtubule will continue to shorten until it disappears or until GTP-tubulin adds back onto the end and a new GTP cap is formed.

For further study, see text section "Dynamic Instability Is an Intrinsic Property of Microtubules," p. 805.

9. Synaptic vesicles added to microtubules did not move along the filaments, even in the presence of ATP. However, the vesicles bound to the microtubules and moved along them after a cytosolic extract from neuronal axons was added to the preparations. The soluble protein (kinesin) in the neuronal extract was purified and characterized using this assay.

For further study, see text section "Kinesin Is a (+) End–Directed Microtubule Motor Protein," p. 812.

10. In addition to the (+) end–directed kinesin, individual axons must contain other motor proteins that are (−) end–directed, such as cytosolic dyneins; the latter are responsible for vesicle motion in the opposite direction.

For further study, see text section "Dynein Is a (−) End–Directed Motor Protein," p. 815.

11. Each member of the kinesin family has a unique tail domain. Because the tail domain is responsible for interacting with specific cargo, each member may transport a different cellular component along microtubules.

For further study, see text section "Each Member of the Kinesin Family Transports a Specific Cargo," p. 815.

12. Demembranated axonemes that have been proteolytically treated to disconnect the doublets from internal structures and radial spokes slide apart when incubated with ATP.

For further study, see text section "Ciliary and Flagellar Beating Are Produced by Controlled Sliding of Outer Doublet Microtubules," p. 820,

13. Individuals with defective dynein produce flagella and cilia that are nonmotile. Because cilia in the trachea are responsible for removing debris, which otherwise could enter the lung, defective cilia can result in coughing and bronchitis. Similarly, movement of sperm cells and transport of eggs through the Fallopian tubes require active flagella and cilia, respectively. Individuals with defective dynein will have nonmotile gametes and thus are sterile.

For further study, see text section "Cilia and Flagella: Structure and Movement," p. 817.

14. The characteristic 9+2 arrangement of axonemes is based (in the case of the 9 outer doublets at least) on the structure of the microtubule bundle termed the basal body. The basal body is actually a set of 9 triplet microtubules but only the complete A tubule and incomplete B tubule seed microtubule assembly (the incomplete C tubule of the basal body does not have a corresponding axoneme microtubule).

For further study, see text section "All Eukaryotic Cilia and Flagella Contain Bundles of Doublet Microtubules," p. 817.

15. Sliding movement is converted to bending movement by cross-linking proteins and inner arm dyneins.

For further study, see text section "Ciliary and Flagellar Beating Are Produced by Controlled Sliding of Outer Doublet Microtubules," p. 820, and "Conversion of Microtubule Sliding into Axonemal Bending Depends on Inner-Arm Dyneins," p. 821.

16. Formation of a bipolar spindle requires centrosome duplication (which occurs prior to M phase) as well as motor proteins that separate the duplicated centrosome to form the two spindle poles. These motors may be present on microtubules in

the overlap zone between the poles and act to push the spindles apart, or they may be on the inner surface of the cell membrane and act to pull astral microtubules and hence the poles apart. It also remains possible that both pushing and pulling forces are needed to separate the poles.

For further study, see text section "Centrosome Duplication Precedes and Is Required for Mitosis," p. 827, and "Organization of the Spindle Poles Orients the Assembly of the Mitotic Apparatus," p. 829.

17. Motor proteins present at the kinetochore may allow this structure to hold onto shortening microtubules. It is not clear whether this activity requires ATP hydrolysis and subsequent force generation by the motor, because kinetochores have been shown to hold onto depolymerizing microtubules in vitro in the absence of ATP.

For further study, see text section "Microtubule Shortening during Anaphase A," p. 831.

18. The separation of spindle poles during anaphase B is very similar to the process of bipolar spindle formation. Motors present on microtubules in the overlap zone between the poles act to push the spindles apart, or motors on the inner surface of the cell membrane act to pull astral microtubules and hence the poles apart. It also remains possible that both pushing and pulling forces are needed to separate the poles during anaphase B. In any case, elongation of microtubules in the overlap zone appears to increase the extent of pole separation.

For further study, see text section "Spindle Elongation during Anaphase B," p. 833.

19. Intermediate filaments provide mechanical support to the plasma membrane and the nuclear membrane and are much more stable than actin filaments or microtubules. This stability is derived from the α helical rod structure of the subunits, which overlap along the filament long axis to generate rope-like filaments.

For further study, see text section "Functions and Structure of Intermediate Filaments Distinguish Them from Other Cytoskeletal Fibers," p. 836.

20. Cancerous cells, particularly at sites of metastasis, often become dedifferentiated, so that a cancer derived from a mesenchymal cell (sarcoma) looks very similar to a cancer derived from an epithelial cell (carcinoma). Thus histological inspection of tumor biopsies may give no clues as to the origin of the tumor, which is important information for design of treatment regimes (chemotherapy vs. radiation vs. surgery, choice of chemotherapy drug, dosage, schedule) and for prognosis. However, many tumor cells still express the IF protein genes characteristic of the cell type from which they arose. For example, an undifferentiated carcinoma would be stained by antibodies specific for keratins, while an undifferentiated sarcoma cell would be stained by antibodies for vimentin. Thus antibody staining for IF proteins allows physicians to diagnose undifferentiated tumors and design appropriate treatment regimes.

For further study, see text section "Intermediate Filaments Can Identify the Cellular Origin of Certain Tumors," p. 838.

21. Intermediate filament subunits do not bind GTP or ATP like tubulin or actin, but instead assembly appears to be regulated by phosphorylation (by kinases) and dephosphorylation (by phosphatases). Addition of a phosphate group to a serine in the N-terminal domain of certain intermediate filament subunits leads to disassembly while removal of the phosphate group allows reassembly.

For further study, see text section "Intermediate Filaments Are Dynamic Polymers in the Cell," p. 840.

22a. If the hand-over-hand model is correct, the one-headed kinesin should not be processive because this protein won't have another head to hold onto the microtubule when the one head releases.

22b. The data in Figure 19-1 demonstrate that two-headed kinesin can support microtubule attachment and movement at lower densities than one-headed kinesin. In addition, the proportional decrease in landing rate as a function of motor density suggests that each two-headed kinesin can support attachment and movement independent of other motors on the coverslip. In comparison, the abrupt decrease in landing rate as a function of motor density for one-headed kinesin suggests that multiple one-headed kinesins are needed to interact with a microtubule and that once below that threshold density value, there will not be enough motors to ensure attachment and movement.

22c. The one-headed kinesin molecule should bind tightly to the microtubule because there is not another head to promote release.

For further study, see text section "Kinesin Is a (+) End–Directed Microtubule Motor Protein," p. 812.

23a. Substitution of the tubulin solution has no effect on microtubule growth while replacement with buffer alone causes microtubule shortening because there is no free tubulin to support elongation. In comparison, the plot shown in Figure 19-2b indicates that even when free tubulin is available, the Kin I motor acts to promote microtubule shortening (although at a rate slower than results from dilution of the tubulin in the chamber).

23b. In the control (Ctrl) and kinesin heavy chain (KHC) samples, all tubulin remained assembled in microtubules as indicated by the presence of tubulin in the pellet fractions. However, Kin I motor in the presence of ATP must have depolymerized the microtubules because the tubulin was found completely in the supernatant fraction (indicating it was no longer assembled). In addition, the Kin I motor must require energy from ATP hydrolysis to produce disassembly because there was no disassembly in the presence of the nonhydrolyzable ATP analog, AMPPNP. In another assay, taxol-stabilized microtubules were incubated alone with ATP, with the Kin I motor and ATP, or with the Kin I motor and AMPPNP, and then centrifuged to pellet tubulin in filaments. Because Kin I can disassemble GMPCPP microtubules, the mechanism by which Kin I causes disassembly is not as simple as promoting hydrolysis of the GTP bound to the subunits that form the GTP cap.

3c. The Kin I motor is probably present at microtubule ends, which are lower in concentration than tubulin subunits, and which are the sites of tubulin addition and loss.

For further study, see text section "Kinesin Is a (+) End–Directed Microtubule Motor Protein," p. 812.

24a. Recovery of fluorescence in a bleached area results from the addition of new, unbleached dimers. Thus the rate of recovery is proportional to the rate of microtubule depolymerization and repolymerization. The $t_{1/2}$ for fluorescence recovery is a few seconds in metaphase cells, whereas it is a few minutes in interphase cells. This difference indicates that microtubules in interphase cells are more stable than those in metaphase cells.

24b. Because most microtubules in a cell are anchored in the MTOC at their (−) ends, it is likely that changes in the (+) ends account for the observed difference in microtubule stability during metaphase and interphase. According to this mechanism, microtubules in metaphase cells must exhibit a greater rate of disassembly at the (+) end than do those in interphase cells. Another possibility is that microtubule-severing proteins are more prominent in interphase cells, because this would give a different pattern of fluorescence recovery than shrinkage and regrowth.

24c. Because X-rhodamine tubulin labels all microtubules in cells, some method was needed to eliminate all microtubules except the kinetochore fibers, which are the focus of the research. Kinetochore microtubules are more stable than other microtubules in the cell, as evidenced by their persistence in a buffer that causes disassembly of other microtubules. Cold depolymerization or colchicine/nocodazole treatment might induce disassembly of most microtubules but leave the kinetochore tubules intact.

24d. Anti-tubulin immunocytochemical analysis of the cell would reveal antibody labeling in the photobleached zone as well as across the rest of the spindle.

24e. Anaphase microtubules move toward the poles, but exhibit no turnover of the microtubules.

24f. These data indicate that metaphase kinetochore tubules turn over, whereas anaphase kinetochore tubules do not.

For further study see text section "Microtubule Dynamics and Motor Proteins During Mitosis," p. 823.

20

Cell-to-Cell Signaling: Hormones and Receptors

PART A: *Chapter Summary*

An elaborate intercellular, communication network coordinates the growth, differentiation, and metabolism of cells in tissues and organs. This chapter discusses the variety of extracellular signaling molecules, receptors, and signal transduction pathways. Communication by extracellular signals usually involves six steps: synthesis and release of the signaling molecule by the signaling cell; transport of the signal to the target cell; detection of the signal by a specific receptor; change in some cellular process in the target cell; and removal of the signal.

Ligand binding to its receptor changes cellular metabolism by a variety of mechanisms. Ligand binding to G protein–coupled receptors activates an associated signal transducing G protein. Depending on the form of the G protein, the activated G protein can either activate or inhibit adenylyl cyclase. Ligand binding to receptor tyrosine kinases activates Ras. In both cases, GTP-GDP plays an important role in the activation/inactivation of G proteins and Ras. Activated Ras induces a kinase cascade that results in the phosphorylation and activation of the kinases Raf, MEK, and MAP kinase. This cascade of sequential reactions provides a huge amplification of an initially small signal.

A number of molecules can act as second messengers in intracellular signaling pathways. These second messengers can be soluble molecules, membrane-bound molecules, or even gases, such as nitric oxide. Second messengers activate certain protein kinases. Cyclic AMP activates cAMP-dependent protein kinases, and the inositol-lipid pathway leads to activation of protein kinase C by diacylglycerol. Calcium bound to calmodulin regulates the activity of many different proteins, including protein kinases.

Signal transduction is a complex process that often involves the activation of more than one signaling pathway. It is the coordinated regulation of these multiple, interacting signaling pathways that ultimately leads to the appropriate activation of transcription factors and modulation of gene expression.

PART B: *Reviewing Concepts*

20.1 *Overview of Extracellular Signaling*
(lecture date _____)

1. What is the difference between endocrine, paracrine, and autocrine signaling?

2. Describe the three broad classes of hormones and the location of their receptors.

3. Briefly describe the four major classes of cell-surface receptors and their modes of action.

20.2 *Identification and Purification of Cell-Surface Receptors*
(lecture date _____)

4. How is the equation for hormone receptor binding similar to the equation for Michaelis-Menten kinetics?

5. Describe how a receptor can be cloned without purifying the receptor protein.

20.3 *G Protein–Coupled Receptors and Their Effectors* (lecture date _____)

6. Describe the experimental approach used to identify functional domains of G-protein coupled receptors.

7. Describe the mechanism of activation of adenylyl cyclase following binding of hormone to a G_s protein-coupled receptor.

20.4 *Receptor Tyrosine Kinases and Ras*
(lecture date _____)

8. Describe the events that occur upon ligand binding to receptor tyrosine kinase receptors.

9. Describe the similarities and differences in the cycling of G_s and Ras between the active and inactive forms.

20.5 *MAP Kinase Pathways*
(lecture date _____)

10. Describe the kinase cascade that transmits signals down from activated Ras protein.

11. Describe how the yeast two-hybrid system was used to search for proteins that interact with Ras.

20.6 *Second Messengers*
(lecture date _____)

12. What is the advantage of having an extracellular signal transmitted by a cascade of sequential events?

13. Describe the mechanism for elevating cytosolic Ca^{2+} via the inositol-lipid signaling pathway.

20.7 *Interaction and Regulation of Signaling Pathways*
(lecture date _____)

14. Describe the regulation of glycogenolysis by different second messengers in muscle cells.

20.8 *From Plasma Membrane to Nucleus*
(lecture date _____)

15. Describe the role of CREB in mediating cAMP control of gene expression

PART C: *Analyzing Experiments*

20.4 *Receptor Tyrosine Kinases and Ras*

16. (••) K. Chou's laboratory has obtained good evidence that the autophosphorylation of tyrosine amino acids in the cytoplasmic domain of the insulin receptor (IR) is essential for insulin action. In their research, Chou and his colleagues have used a variety of techniques including site-directed mutagenesis of the insulin receptor and introduction of antibodies to the IR kinase domain into whole cells. In the case of site-directed mutagenesis, IR genes were created in which a lysine codon in the active site of the IR kinase domain was converted to an alanine. ATP could not bind to the active site of these mutant receptors, and autophosphorylation of the receptor was prevented. When these mutant receptors were expressed in Chinese hamster ovary (CHO) cells, they were found to bind insulin normally. The transfected CHO cells had ~20,000 mutant receptors per cell. When these transfected cells were treated with insulin, no enhancement in insulin-stimulated parameters (e.g., glycogen synthesis, thymidine incorporation into DNA, and S6 kinase activation) was detected compared with nontransfected control cells expressing ~2000 wild-type receptors per cell, as shown in Figure 20-1 (bar TM vs. C).

Other investigators incorporated antibodies against the IR cytoplasmic tyrosine kinase domain into liposomes. When cells are exposed to these liposomes, the antibodies can gain access to the cytoplasmic surface of the plasma membrane. Treatment of transfected CHO cells expressing ≈20,000 wild-type receptors per cell with these antibody-laden liposomes depressed insulin-stimulated thymidine incorporation below that of control nontransfected cells expressing ≈2000 wild-type receptors per cell (see Figure 20-1, bar TW + AB vs C). In contrast, thymidine incorporation by transfected CHO cells expressing ≈20,000 wild-type receptors per cell was much greater than in the control nontransfected cells. Assume that all cells in each sample expressed the receptor types indicated.

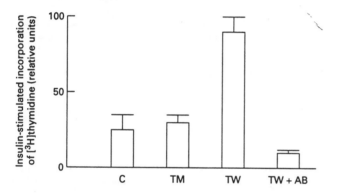

C = control, nontransfected CHO cells
    with wild-type receptors (≈ 2000/cell)
TM = cells transfected with mutant receptors (≈ 20,000/cell)
TW = cells transfected with wild-type receptors (≈ 20,000/cell)
TW + AB = cells transfected with wild-type receptors + liposome-added
    antibodies against tyrosine kinase domain of insulin receptor

Figure 20-1

a. Assuming that the cell surface receptors in all four samples have a similar affinity for insulin (i.e., similar K_D values), why is insulin-stimulated thymidine incorporation lower in the TW + AB sample than in the C sample but the activity of the TM sample is the same as the control?

b. When the phosphorylated cytosolic proteins produced during incubation of the TW cells with insulin were analyzed by two-dimensional gel electrophoresis, only one phosphorylated protein was detected. Since most polypeptide hormones cause an increase in the phosphorylation of several cytosolic and/or particulate proteins, does this finding indicate that the enhanced thymidine in-

corporation observed in TW cells does not represent an enhanced insulin-specific stimulation of insulin receptors?

20.6 *Second Messengers,*
20.7 *Interaction and Regulation of Signaling Pathways*

17. (•••) Considerable effort has been expended on determining how different hormones can elicit the same physiological response in cells. It is now known that many receptors share the same pool of adenylate cyclase, thus integrating responses at the level of the plasma membrane. For example, in a celebrated study of this type, it was noted that both nerve growth factor (NGF) and stimulators of adenylate cyclase (dbcAMP) can induce differentiation of the transformed cell line PC12 into a cell resembling a sympathetic neuron. Other reports have indicated that NGF can stimulate the accumulation of cAMP, implying that NGF acts through the cAMP pathway. However, most of the available evidence indicates that the NGF and cAMP pathways are distinct from each other.

C. Richter-Landsberg and B. Jastorff have used two cAMP analogs, which act as either an agonist or antagonist of the cAMP-dependent protein kinases, to further examine the possible overlap of the cAMP and NGF pathways. The agonist is (Sp)-cAMPS and the antagonist is (Rp)-cAMPS. These workers incubated PC12 cells with combinations of NGF and these analogs and then determined the percentage of cells with neurites, which are a hallmark of differentiated PC12 cells. Thus an increase in the percentage of cells with neurites is a direct, morphological indication of differentiation. The results of this experiment are presented in Figure 20-2.

a. What do the data in Figure 20-2 indicate about the relationship between the NGF and cAMP pathways in these cells?

b. In subsequent experiments, this group monitored the differentiation of PC12 cells in the presence of forskolin, a stimulator of adenylate cyclase, and in the presence of forskolin + saturating levels of NGF, as illustrated in Figure 20-3. How do these data affect your answer to part (a)?

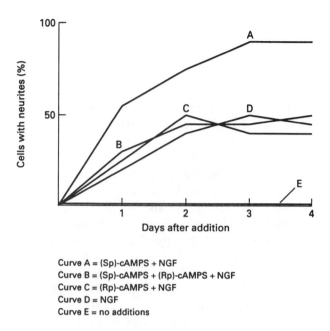

Curve A = (Sp)-cAMPS + NGF
Curve B = (Sp)-cAMPS + (Rp)-cAMPS + NGF
Curve C = (Rp)-cAMPS + NGF
Curve D = NGF
Curve E = no additions

Figure 20-2

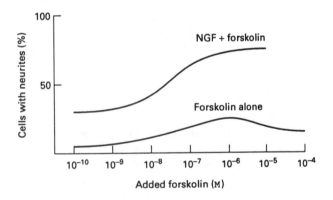

Figure 20-3

18. (••) Many experimental approaches can be used to determine if two or more different second messengers can regulate the same biosynthetic process. For example, specific agonists and antagonists can be used to facilitate dissection of signal-transduction pathways in cells, as described in problem 17. However, highly specific antagonists to all hormones and second-messenger systems are not yet available and alternative approaches must be used in some cases. One such situation involves the ability of NGF and protein kinase C to activate ornithine decarboxylase (ODC) in PC12 cells. As noted in problem 17, NGF can induce differentiation of these cells. Ornithine de-

carboxylase (ODC) is a key regulatory enzyme in the production of polyamines. Although the latter are associated with cell differentiation, they may not be obligatory to the NGF-induced differentiation of PC12 cells. Nonetheless, ODC is a marker of differentiation in this cell line and is induced by both NGF and protein kinase C.

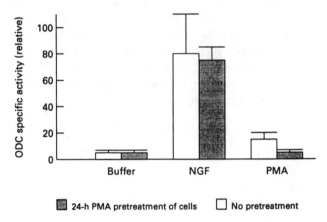

■ 24-h PMA pretreatment of cells   □ No pretreatment

Figure 20-4

a. In an attempt to sort out the pathways by which NGF and protein kinase C activate ODC, PC12 cells were pretreated with the tumor promoter PMA for 24 h to down-regulate protein kinase C. Next, aliquots of pretreated and untreated PC12 cells were incubated with buffer alone, NGF, or PMA and then the activity of ODC was measured. The results are shown in Figure 20-4. Do these data indicate that the activation of ODC by protein kinase C and NGF occurs by the same or different pathways?

b. L. Greene of Columbia University has created a mutant PC12 cell line that has no high-affinity NGF receptors. When presented with NGF, these cells do not increase their ODC activity compared with uninduced control levels. Suppose that samples of these mutant cells are incubated in the presence (+) and absence (–) of PMA and NGF, as indicated in Figure 20-5, and that the ODC activity is then determined. Based on your conclusion in part (a) and the partial results shown in Figure 20-5, predict the levels of ODC activity for the PMA⁻/NGF⁺ and PMA⁺/NGF⁻ samples, which are not shown in the Figure.

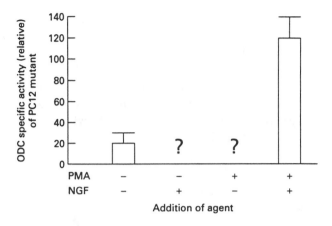

Figure 20-5

20.4 Receptor Tyrosine Kinases and Ras

19. (•••) Platelet-derived growth factor (PDGF) is a growth-promoting substance that is released from platelets when they adhere to the surface of an injured blood vessel. Binding of PDGF to fibroblasts and other cell types produces extremely diverse effects including an increase in DNA synthesis, changes in ion fluxes, alterations in cell shape, and changes in phospholipid metabolism. These effects are mediated by the PDGF receptor, a receptor tyrosine kinase (RTK). After activation and autophosphorylation of the PDGF receptor, phosphotyrosines in its cytosolic domain can interact with SH2 domains in several different cytosolic proteins, leading to various downstream effects. The cytosolic domain of the PDGF receptor has four binding sites, each containing one or two tyrosine residues; when these are phosphorylated, they can bind PI-3 kinase, GAP, the γ isoform of phospholipase (PLC$_\gamma$) and Syp (a protein phosphatase).

A. Kazlauskas and co-workers were interested in the activation of specific signaling pathways by PDGF. In order to understand which of these multiple binding sites are responsible for activating specific responses, they created a series of mutant-gene-encoding PDGF receptors in which tyrosine residues (Y) were replaced by nonphosphorylatable phenylalanine residues (F) in three of these binding sites, as diagrammed in Figure 20-6. These mutant receptors should bind to only one of the four possible cytosolic proteins, and theoretically should activate only one of the four possible signaling pathways. For example, the mutant receptor known as Y40/51 contains phe-

nylalanine residues at positions 771, 1009, and 1021. Such a receptor should interact specifically only with PI-3 kinase, which is known to bind to phosphotyrosine residues at position 40 and 51. The F5 receptor contains phenylalanine replacements for all the phosphorylatable tyrosine residues, and should therefore not interact with any of the intracellular signaling proteins. Take a minute to examine Figure 20-6 carefully to ensure that you understand this concept.

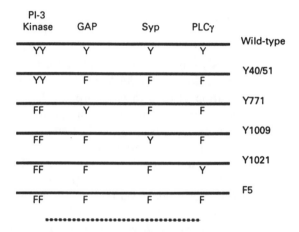

PI-3 Kinase	GAP	Syp	PLCγ	
YY	Y	Y	Y	Wild-type
YY	F	F	F	Y40/51
FF	Y	F	F	Y771
FF	F	Y	F	Y1009
FF	F	F	Y	Y1021
FF	F	F	F	F5

Cytosolic domain of PDGF receptor

Figure 20-6

a. These researchers transfected HepG2 cells, which do not produce an endogenous PDGF receptor, with the genes encoding these mutant receptors. Appropriate controls indicated that the mutant receptors expressed in the transfected cells did indeed associate with the appropriate cytosolic proteins. They then assayed PDGF-induced [³H]-thymidine incorporation into DNA in these transfected cell lines. In this assay, cells were plated at a subconfluent cell density, arrested by serum deprivation, and then stimulated with various concentrations of PDGF. After 18–20 h, the cells were pulsed for 2 h with [³H]-thymidine, and the amount of incorporated radioactivity was measured by scintillation counting of TCA-precipitated nucleic acids. The results, shown in Figure 20-7 are expressed as the percentage of the response stimulated by serum. Based on the results shown in this Figure, which intracellular signaling pathways are involved in initiation of DNA synthesis in response to PDGF?

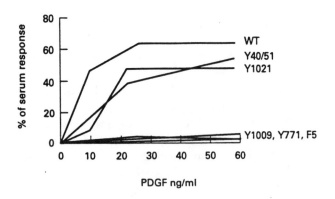

Figure 20-7

b. Previously other workers found that activation of a protein known as S6 kinase is important in stimulation of cells to pass through the $G_1 \rightarrow S$ transition in the cell cycle. This protein, which exists in two forms designated αI and αII, is activated by phosphorylation of serine residues, and thus is not a direct substrate for the PDGF receptor tyrosine kinase. Receptor tyrosine kinases activate MAP kinase via the Sos and c-Ras pathway, but it is not known which pathway(s) is coupled to phosphorylation of the S6 kinase. In order to answer this question, Kazlauskas and coworkers generated another series of PDGF receptor mutants. In these mutants, most of the tyrosines were retained, but specific tyrosines were changed to phenylalanines. For example, a mutant receptor with a phenylalanine residue at position 1021 was generated and designated F1021. The researchers incubated quiescent HepG2 cells, expressing these various mutant PDGF receptors, with PDGF for 45 minutes, and then assayed S6 kinase activity and extent of phosphorylation. Figure 20-8 shows stimulation of S6 kinase activity in response to PDGF in these cells, expressed as times the activity in untreated cells. Figure 20-9 shows a Western blot of electrophoretically separated S6 kinase from these cells in the presence and absence of PDGF. Phosphorylation of the 70-kD (αI) and 85-kD (αII) forms of S6 kinase results in species that migrate more slowly on SDS-PAGE gels. In this figure the direction of electrophoretic movement is from top to bottom. Based on these data, which intracellular signaling pathways are involved in phosphorylation and activation of S6 kinase?

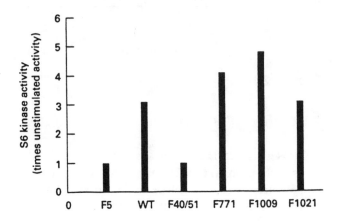

Figure 20-8

20. (••) Epidermal growth factor (EGF) is a mitogen for many different cell types. In some cells the mitogenic action of EGF is associated with elevation of the intrinsic tyrosine kinase activity of the EGF receptor, which results in phosphorylation of tyrosine residues on the receptor (autophosphorylation) and on many other proteins. Receptor autophosphorylation is thought to occur after dimerization of two EGF receptors, with each activated receptor subunit phosphorylating the other. The activated receptor can also phosphorylate various cytosolic proteins.

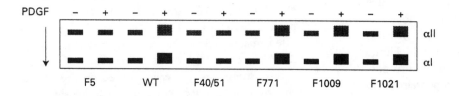

Figure 20-9

Activation of the tyrosine kinase activity of the EGF receptor and subsequent receptor autophosphorylation are thought to be required for the EGF-induced mitogenic response in receptive cells.

Recently, however, A. Ullrich and co-workers discovered that although EGF is a potent mitogen for a rat mammary adenocarcinoma cell known as MTLn3, the receptor in these cells is *not* autophosphorylated in response to EGF. Further investigation of this phenomenon indicated that these cells expressed a complete form of the rat EGF receptor, containing all the sequences that normally are phosphorylated in response to EGF. Binding experiments indicated that these cells exhibited both high-affinity ($K_D = 0.17$ nM) and low-affinity ($K_D = 1.2$ nM) binding sites for EGF; they possess $\approx 56,000$ receptors per cell. Experiments involving chemical cross-linking of iodinated EGF to the cell surface showed that EGF bound specifically to a 170- to 180-kD protein; this protein cross-reacted with antisera specific for the intracellular domain of the EGF receptor. At this point these researchers concluded that the MTLn3 cells had structurally normal EGF receptors.

a. In order to demonstrate that the EGF receptor in MTLn3 cells is active, cell lysates from EGF-treated or untreated cells, grown in the presence of [^{32}P]-labeled orthophosphate, were immunoprecipitated with antibodies specific for phosphotyrosine; the precipitated proteins then were separated by electrophoresis. A positive control was provided by human A431 cells, which overexpress the EGF receptor. These results are shown in Figure 20-10. The legend at the left indicates the approximate molecular weights of the separated proteins. After electrophoresis, the gel was treated with KOH to remove phosphate groups from serine and threonine residues on these proteins, assuring that only radioactivity in phosphotyrosine was retained on the gels. What is your interpretation of the results shown in Figure 20-10?

b. One possible explanation of the lack of autophosphorylation of the EGF receptor in intact MTLn3 cells is that the membrane-bound receptor is incapable of dimerization, which is thought to allow phosphorylation of one receptor subunit by another. To test this hypothesis, Ullrich and co-workers treated adherent MTLn3 and A431 cells with a low concentration of nonionic detergent (0.15% Triton X-100) for 1 min. Previously they had shown that this mild detergent treatment removes most of the membrane constituents, leaving cell cytoskeletal elements and associated membrane proteins. They then incubated these detergent-treated cells with radioactive ATP in the presence and absence of 150 ng/ml EGF for 10 min. Electrophoretic separation and autoradiography were performed to analyze the labeled phosphorylated proteins under these conditions; the results are shown in Figure 20-11. What is your interpretation of these results?

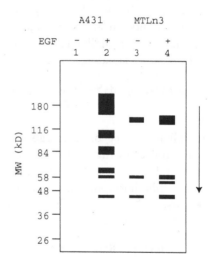

Figure 20-10

c. Current dogma holds that activation of the EGF receptor kinase is causally linked to dimerization of receptors. This conclusion is based primarily on the results of experiments performed in the presence of detergent. Do the results shown in Figures 20-10 and 20-11 support this hypothesis? Why or why not?

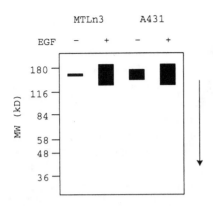

MTLn3   A431

EGF   –   +   –   +

Figure 20-11

Answers

1. In endocrine signaling, signaling molecules are synthesized by one organ and are carried to target cells by the blood. In paracrine signaling, the signaling molecules are released and affect only target cells in close proximity. In autocrine signaling, the cell that releases the signaling molecule is also affected by the released signaling molecule.

For further study, see text, pp. 849–850 and Figure 20-1.

2. Most hormones fall into three basic categories: 1. small lipophilic molecules that diffuse across the cell membrane and interact with intracellular receptors; 2. hydrophilic hormones that bind to cell surface receptors; and 3. lipophilic hormones that bind to cell surface receptors.

For further study, see text, pp. 850–852 and Figure 20-2.

3. The four major classes of cell surface receptors include: G protein coupled receptors, ion channel receptors, tyrosine kinase linked receptors, and receptors with intrinisc enzymatic activities. For G-protein coupled receptors, ligand binding activates a G protein, which in turn activates or inhibits an enzyme that synthesizes a specific second messenger. For ion-channel receptors, ligand binding changes the conformation of the receptor allowing ions to flow through it. For tyrosine kinase-linked receptors, ligand binding stimulates dimer formation and activation of a cytosolic protein kinase. For receptors with intrinisic enzymatic activity, ligand binding activates an enzymatic activity e.g., kinase or phosphatase.

For further study, see text, pp. 852–854 and Figure 20-3.

4. From the equation for hormone-receptor binding, [RH]/R_T, which is a measure of the amount of hormone complexed with receptor divided by the sum of the free and bound receptor is comparable to V_{max} for enzyme kinetics. Also K_D describes the dissociation constant of the receptor-hormone complex, which is comparable to the K_m for enzyme kinetics. The Kd and Km both measure the affinity of the receptor for its ligand or the enzyme for its ligand.

For further study, see text, pp. 858–859.

5. To clone a receptor without purifying the receptor protein, a plasmid cDNA library is first prepared from cells that synthesize the receptor. This cDNA library is screened by transfecting the cloned cDNAs into cells that do not express the receptor. Cells that take up the cDNA encoding the receptor are detected by their ability to bind radiolabeled or fluorescent ligand. The cDNA encoding the receptor can then be isolated from the transfected cells.

For further study, see text, pp. 860-861 and Figure 20-9.

6. The functional domains of G-protein coupled receptors were determined by experiments using recombinant chimeric receptor proteins containing parts of the β_2 and α_2 adrenergic receptors. These chimeric receptors were tested for their ligand binding specificity and their ability to activate or inhibit adenylate cyclase. The results of these studies demonstrated that α helix 7 and the C-terminal domain of the receptor play a role in determining ligand binding specificity and the cytosolic loop between α helices 5 and 6 interacts with G proteins.

For further study, see text, pp. 863–866 and Figure 20-14.

7. Binding of hormone to its receptor causes a conformational change in the receptor. The receptor then binds to $G_{s\alpha}$ protein causing a conformational change in the $G_{s\alpha}$ protein and release of the bound GDP. GTP then binds to $G_{s\alpha}$, which causes release of the GTP-$G_{s\alpha}$ complex from the β and γ

subunits. The GTP-$G_{s\alpha}$ complex then activates adenylyl cyclase. Activation is short-lived because the intrinisic GTPase activity in $G_{s\alpha}$ hydrolyzes GTP to GDP and returns $G_{s\alpha}$ to its inactive form

For further study, see Animation 20.2 "Extracellular Signaling," on the MCB 4.0 CD-ROM, as well as text, pp. 865–868 and Figure 20-16.

8. Ligand binding to receptor tyrosine kinases causes a conformational change in receptor monomers that promotes dimerization. The intrinisic protein kinase activity of each receptor subunit phosphorylates tyrosine residues near the catalytic site in the other subunit. In subsequent reactions, tyrosine residues in other parts of the receptor molecule are autophosphorylated.

For further study, see text, p. 872 and Figure 20-21.

9 For both G_s and Ras proteins, the G_s- and Ras-GDP complex represents the inactive form. After ligand binding to the appropriate receptor, GDP is removed from G_s and Ras and is replaced by GTP, which represents the active form. In the case of G_s the intrinsic GTPase activity of G_s converts GTP to GDP and returns Gs to its inactive form. In the case of Ras, two additional proteins play a role in Ras cycling. Guanine nucleotide exchange factor (GEF) binds to Ras-GDP and causes release of GDP. GTP then binds to Ras activating it. Active Ras with GTP bound is cycled to inactive Ras by the GTPase activating protein (GAP).

For further study, see text, pp. 872–873 and Figures 20-16 and 20-22.

10. Activated Ras binds to the serine/threonine kinase, Raf. Raf binds to and phosphorylates MEK, a kinase that phosphorylates both tyrosine and serine residues. MEK phosphorylates and activates MAP kinase, another serine/threonine kinase. MAP kinase then phosphorylates many different target proteins.

For further study, see text, pp. 878–879 and Figure 20-28.

11. The yeast two-hybrid system was used for detecting proteins that interact with Ras. Two plasmids were constructed. One plasmid contained the Ras gene fused to a DNA binding domain and the other plasmid contained an activation domain fused to a cDNA library. These plasmids were co-transfected into yeast cells with mutations in genes required for tryptophan, leucine, and histidine biosynthesis and then grown in the absence of tryptophan and leucine. Only cells that synthesize a protein that interacts with Ras would grow, thus identifying cDNAs encoding Ras-binding proteins.

For further study, see Animation 20.1 "Yeast Two-Hybrid System" on the MCB 4.0 CD-ROM, as well as text, pp. 878–880 and Figure 20-29.

12. Transduction of an extracellular signal via a cascade of sequential events is advantageous because of the amplification of the signal. The binding of a few molecules to a receptor can result in the synthesis of a large number of effector molecules because at each step of the cascade the signal is enzymatically amplified.

For further study, see text, pp. 886–887 and Figure 20-37.

13. Binding of a hormone to its receptor leads to activation of G protein, which in turn activates phospholipase C. Phospholipase C then cleaves phosphatidylinositol 4,5-bisphosphate (PIP_2) to inositol 1,4,5-trisphosphate (IP_3) and diacylglycerol (DAG). The IP_3 diffuses through the cytosol and interacts with IP_3 sensitive Ca^{2+} channels in the membrane of the endoplasmic reticulum causing release of Ca^{2+} stores.

For further study, see Animation 20.3 "Second Messengers in Signaling Pathways" on the MCB 4.0 CD-ROM, as well as text, pp. 889–890 and Figure 20-39.

14. Stimulation of muscle cells by nerve impulses causes an increase in cytosolic Ca^{2+} concentrations. This rise in Ca^{2+} concentration activates glycogen phosphorylase kinase (GPK), which in turn activates glycogen synthase by phosphorylating it. In addition epinephrine binding to β-adrenergic receptors leads to increased cytosolic concentrations of Ca^{2+} and cAMP. GPK is also activated by a cAMP-dependent protein kinase. In this way, the two second messengers, Ca^{2+} and cAMP, can regulate the same metabolic pathway.

For further study, see text, p. 897 and Figure 20-44.

15. Genes regulated by cAMP contain a cis-acting element called the cAMP response element (CRE). In response to an increase in cAMP, a cAMP dependent protein kinase phosphorylates the transcription factor called CRE-binding protein (CREB). Phosphorylated CREB protein binds to the CRE element and stimulates transcription in conjunction with the co-activator CBP/P300.

For further study, see Animation 20.2 "Extracellular Signaling" on the MCB 4.0 CD-ROM, as well as text, pp. 902–903 and Figure 20-48.

16a. The insulin-stimulated thymidine incorporation in the control cells is mediated by the endogenous insulin receptors ($\approx$2000/cell). Expression of mutant receptors in transfected CHO cells (TM sample) does not affect the activity of these native insulin receptors. However, when the tyrosine kinase antibodies are introduced into the TW + AB sample, the antibodies depress the tyrosine kinase activity of both the endogenous receptors and the transfected wild-type receptors present in these cells. As a result, insulin-stimulated thymidine incorporation in the TW + AB sample is below that of the control cells.

16b. No. Very few endogenous cytosolic proteins that are phosphorylated by insulin have been identified. Therefore the single phosphorylated protein on the gel is most probably the phosphorylated form of the insulin receptor itself. Although increasing the receptor number through transfection should increase the chances of detecting phosphorylated cytosolic proteins, which are in low concentration in the cell, the inability to do so has nothing to do with the validity of the data presented.

17a. The inability of the cAMP antagonist to block the effect of NGF [(Rp)-cAMPS + NGF curve] and the additive effect of the cAMP agonist and NGF [(Sp)-cAMPS + NGF curve] both suggest that NGF does not act through the cAMP pathway. In other words, two separate pathways exist to mediate the effects of NGF and cAMP.

17b. The data in Figure 20-3 suggest that NGF and cAMP may have a synergistic effect. The forskolin-only curve indicates that at concentrations below 10^{-8} M, forskolin has little stimulatory effect on differentiation of PC12 cells; the maximum forskolin effect of 20% occurs at about 10^{-6} M. At low forskolin concentrations ($<10^{-8}$ M), the NGF + forskolin curve represents the maximal activity of NGF, which is present at saturating levels. If NGF and forskolin had merely additive effects, then the maximal combined effect should have been about 50% (30% from NGF + 20% from forskolin) at 10^{-6} M. Since the observed combined effect at this concentration was 70%, NGF and forskolin probably are acting synergistically.

18a. The data in Figure 20-4 suggest that ODC activation is mediated by two separate pathways. Pretreatment with PMA, which causes downregulation of protein kinase C, has no effect on the basal level of ODC activity in buffer but eliminates the ability of PC12 cells to increase their ODC activity in the presence of PMA. In contrast, NGF induction of ODC activity was unaffected by pre-treatment with PMA, which down-regulated protein kinase C.

18b. Since the mutant cells lack NGF receptors and do not exhibit NGF induction of ODC activity, the activity in the PMA$^+$/NGF$^+$ sample represents PMA induction only. Therefore, the PMA$^+$/NGF$^-$ sample should be similar to the PMA$^+$/NGF$^+$; in both cases only PMA induction is occurring. The PMA$^-$/NGF$^+$ sample should be similar to the PMA$^-$/NGF$^-$ sample, which exhibits only the basal uninduced ODC activity.

19a. These results indicate that PDGF receptors coupled to the PI-3 kinase pathway (Y40/51 curve) and PLC$_\gamma$ pathway (Y1021 curve) initiated near wild-type levels of DNA synthesis. Binding of GAP or Syp did not initiate DNA synthesis. These findings indicate the presence of redundant multiple signaling pathways in PDGF-induced mitogenesis.

19b. All the mutant receptors with partial substitution of phenylalanine for tyrosine exhibit normal phosphorylation (Figure 20-9) and activation (Figure 20-8) of S6 kinase except F40/51. These findings indicate that the tyrosine residues at positions 40 and 51 of the PDGF receptor are required for both phosphorylation and activation of the S6 kinase. Since these are the residues involved in binding PI-3 kinase (see Figure 20-6), phosphorylation and activation of S6 kinase is linked to the PI-3 kinase pathway.

20a. EGF treatment of MTLn3 cells results in phosphorylation of tyrosine residues on several cytosolic proteins, as indicated by the strong bands in lane 4 at molecular weights less than 180 kD, which are missing (or less intense) in untreated cells (lane3). However, there is no band corresponding to the phosphorylated EGF receptor in lane 4, whereas the phosphorylated receptor ($\approx$180-kD band) is found in the A431 cells (lane 2), as expected. Thus EGF seems to stimulate the kinase activity of the receptor in MTLn3 cells, but this activation does not result in autophosphorylation of membrane-bound receptor, as it does in the A431 cells.

20b. There is an increase in the amount of phosphorylated EGF receptor (at 170–180 kD) in both cell types in the presence of EGF. This finding indicates that the EGF receptor in the MTLn3 cells is fully capable of autophosphorylation under certain conditions where membrane components are removed. It is likely that the EGF receptor is aggregated and bound to the cytoskeletal network after mild detergent treatment, thus one activated receptor monomer can phosphorylate another under these conditions.

20c. These results indicate that dimerization is not required for activation of the tyrosine kinase activity of the EGF receptor, since EGF-dependent tyrosine phosphorylation can be detected in the absence of receptor phosphorylation, as indicated by the data in Figure 20-10. Furthermore, receptor phosphorylation is not required for a mitogenic response to EGF in MTLn3 cells. Although it is abundantly clear that receptor dimerization and subsequent autophosphorylation do occur in most cell types in response to EGF, these data indicate that receptor kinase activation is an independent phenomenon, which is not causally linked to the dimerization or phosphorylation status of the EGF receptor.

Additionally, closer examination of these figures reveals another often ignored phenomenon. Note the very low basal autophosphorylation activity (in the absence of EGF) of the membrane-bound EGF receptor in intact A431 cells in Figure 20-10, and compare it with the significantly elevated basal activity of the receptor in detergent-treated cells in Figure 20-11. Detergent treatment alone seems to activate the receptor in this case. These results should make one cautious about extrapolation of results obtained from detergent-treated membrane proteins; under these conditions the proteins may not be structurally or functionally identical to their membrane-bound counterparts.

21 Nerve Cells

PART A: *Chapter Summary*

Neurons are specialized for receiving, conducting, and delivering signals in multicellular organisms. Each neuron is able to conduct information along its length in the form of a propagating electrical disturbance termed the action potential, and neurons pass signals to other neurons or electrically excitable cells (such as muscle cells) across a narrow gap termed the synapse. Propagation of the action potential depends on voltage-gated ion channels in the neuron membrane, while communication across the synapse typically involves the release by one neuron of soluble signaling molecules termed *neurotransmitters* and the detection of these molecules by the downstream neuron or muscle cell. Although researchers have a reasonable understanding of how isolated neurons function, the huge number of neurons in the brain (10^{12} in the human brain) and the large number of connections (1000's) that each neuron has with other neurons has so far made understanding the higher level processes such as learning and memory very difficult.

PART B: *Reviewing Concepts*

21.1 *Overview of Neuron Structure and Function* (lecture date _____)

1. What is a reflex arc? What constitutes the central nervous system? The peripheral nervous system?

2. Compare the roles of dendrites and axons.

3. Compare the roles of the action potential and synapse in nerve cell signaling.

21.2 *The Action Potential and Conduction of Electric Impulses* (lecture date _____)

4. Which property of the voltage-gated Na^+ channel ensures that action potentials will only be propagated in one direction?

5. Explain why myelination of an axon increases the speed at which it can propagate action potentials.

6. If the external Na^+ concentration is decreased, the magnitude of the action potential in a squid giant axon is decreased. Why? What would happen if the external Na^+ concentration were reduced to zero?

7. Fibroblasts and most other non-neuronal cells exhibit an inside-negative electric potential. However, when they are depolarized, fibroblasts do not produce an action potential even though the concentrations of Na^+ and K^+ inside and outside fibroblasts are identical to those associated with neurons. Why do fibroblasts not generate an action potential?

8. Changes in the ion permeability of the membrane of a neuron alter the membrane potential of the cell. Which of the following would lead to a depolarization of the neuron membrane, and which would lead to a hyperpolarization?

   a. increase in K^+ permeability

   b. decrease in Cl^- permeability

   c. increase in Na^+ permeability

   d. decrease in K^+ permeability

21.3 *Molecular Properties of Voltage-Gated Ion Channels* (lecture date _____)

9. Many voltage-sensitive channel proteins contain several transmembrane domains. Typically, one membrane-spanning segment in each domain is an α helix with lysine or arginine at approximately every third or fourth position.

   a. What is the proposed function for these conserved α-helical segments, and how are they thought to work on a molecular level?

   b. Describe experimental evidence that supports this function for the conserved α helices.

10. *Drosophila shaker* mutants have a defective K⁺ channel that causes delayed re-polarization of the plasma membrane of axons. However, this type of delay also can result from other membrane-specific defects. Suggest one other possible defect that would contribute to delayed re-polarization of neurons. Which techniques are most suitable for demonstrating such a defect?

11. How are *Xenopus* oocytes used to aid in the study of voltage-gated ion channels?

21.4 *Neurotransmitters, Synapses, and Impulse Transmission* (lecture date _____)

12. How does impulse transmission at inhibitory synapses differ from that at excitatory synapses?

13. The ability to distinguish electric and chemical synapses is critical in determining the function of a particular neural circuit. Describe two experimental approaches for differentiating the two types of synapses.

14. What is the function of the voltage-gated Ca²⁺ channels that are present in the neuronal membrane around axon terminals? What would happen to synaptic transmission if the pre- and postsynaptic cells were incubated in Ca²⁺-free medium?

15. Termination of synaptic signaling requires that neurotransmitters be removed from the synaptic cleft. How is this accomplished?

21.5 *Neurotransmitter Receptors* (lecture date _____)

16. The current model of the nicotinic acetylcholine receptor molecule has five subunits arranged around a central channel, or pore, whose diameter is slightly less than 1 nm, which is much larger than the dimensions of Na⁺ or K⁺ ions. Why is the channel pore so large compared with the size of the ions that pass through it?

17. Acetylcholine stimulates skeletal muscle cells to contract but slows the rate of contraction of cardiac muscle cells. How is this possible?

18. Why are GABA and glycine considered inhibitory neurotransmitters?

21.6 *Sensory Transduction* (lecture date _____)

19. Compare the detection mechanisms for taste and olfaction. What is unique about olfactory neurons?

20. Explain why stimulation of the rod photoreceptors of the eye can be considered the "reverse" of a typical neuron. When a rod cell is injected with cGMP, the synaptic activity of the cell increases. Explain this effect in terms of the properties of the Na⁺ channels in the rod-cell membrane.

21.7 *Learning and Memory* (lecture date _____)

21. Why is the sea slug Aplysia a useful model for short-term learning?

PART C: *Analyzing Experiments*

21.2 *The Action Potential and Conduction of Electric Impulses*

22. (•) Much of our understanding of nerve cell signaling is due to the ability to measure changes in the membrane potential of isolated neurons (see text Figure 21-7, p. 917). Below are three different examples of experiments based on this approach

   a. The simple nervous system of the leech, an invertebrate, has been extensively studied, and many of its component circuits are well characterized. As illustrated in Figure 21-1a, one such circuit consists of three presynaptic neurons (A, B, and C) that synapse with a postsynaptic neuron. *In situ* analyses indicate that each presynaptic neuron contributes an excitatory component to the postsynaptic cell. Tracings of the changes in membrane potential following stimulation of the presynaptic cells are shown in Figure 21-1b; the changes were measured with intracellular microelectrodes located at the arrows in Figure 21-1a. When neuron A or B is inhibited using voltage-clamping techniques, the postsynaptic cell does not fire. However, if neuron C is inhibited, stimulation of neurons A and B is sufficient to produce an action potential in the postsynaptic cell. Based on the data in Figure 21-1b, suggest the most plausible reason for this unexpected finding.

b. When a neurotoxin isolated from a mud-dwelling fish from the Amazon River is placed in the solution bathing an isolated neuron, it affects the action potential of the neuron as shown in Figure 21-1c. What is the probable mechanism of action of this drug on this neuron?

c. A certain neurotoxin depolarizes the resting potential and diminishes the action potential of a neuron as shown in Figure 21-1d. How could you re-create the effect of this toxin simply by manipulating the ion concentrations surrounding the neuron in question?

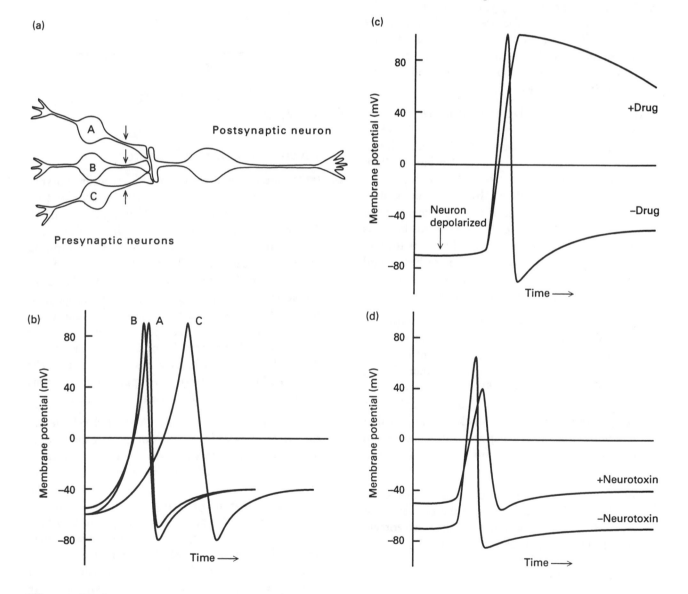

Figure 21-1

21.4 *Neurotransmitters, Synapses, and Impulse Transmission*

23. (•••) Synapsin I is a major neuron-specific phosphoprotein located on the cytoplasmic surface of synaptic vesicles. It is possible to separate synapsin I from isolated vesicles and then to reconstitute synaptic vesicles by incubating free synapsin I with the stripped experimental vesicles under various conditions.

   a. What can you conclude from the data in Figure 21-2 about the binding of synapsin I to synaptic vesicles?

   b. In a similar experiment, protease-treated vesicles were unable to incorporate synapsin I with the same affinity as non-protease-treated vesicles. What do these data and the high-salt curve in Figure 21-2 suggest about the formation of these reconstituted vesicles?

   c. Synapsin I has a collagenase-insensitive head domain, which can be phosphorylated by cAMP-dependent protein kinase or Ca^{2+}-calmodulin–dependent protein kinase I. It also has an elongated collagenase-sensitive tail domain, which can be phosphorylated by Ca^{2+}-calmodulin–dependent protein kinase II. Experiments with the isolated tails and heads have shown that tails bind less well to protease-treated vesicles than to non-protease-treated vesicles, whereas heads bind equally well to both types of vesicles. What is a possible interpretation of these results?

   d. In other experiments, a hydrophobic photoaffinity label was used to label proteins in synaptic vesicles. This probe can gain entry into the hydrophobic domain of proteins. The same labeling pattern was found (using SDS gel electrophoresis and autoradiography) with both endogenous vesicles and reconstituted vesicles (i.e., those that had been stripped of synapsin I and reconstituted by exposure to free synapsin I). What is the significance of this experiment?

   e. What type of experiment could be performed to determine whether one or more of the protein bands revealed in the photoaffinity-labeling experiment described in part (d) is, indeed, synapsin I?

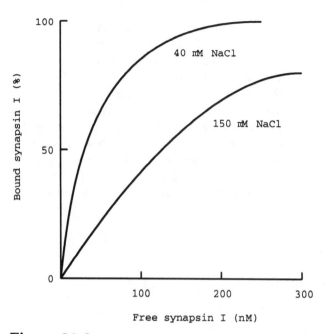

Figure 21-2

21.5 *Neurotransmitter Receptors*

24. (•••) One of the three glutamate receptor subtypes is a ligand-gated Ca^{2+} channel that is activated by glutamate as well as N-methyl-D-aspartate (NMDA). This receptor may be involved in the death of hippocampal cells in the central nervous system during episodes of cerebral ischemia (lack of O_2 to the brain). According to this hypothesis, the large increase in extracellular glutamate in the hippocampus that occurs during cerebral ischemia overstimulates these NMDA receptors, leading to a large influx of Ca^{2+} ions into hippocampal neurons and their subsequent death. This mechanism is thought to be the primary reason why hippocampal neurons are among the first to die during cerebral ischemia.

In order to investigate NMDA-elicited cell death, the effect of NMDA on the survival of embryonic hippocampal cells in vitro was determined. In these studies, hippocampal neurons were isolated from 18-day-old rat embryos and placed in culture; the number of neurons surviving was determined periodically over a 2-week period using a trypan blue exclusion test. Among other things, researchers analyzed the ability of MK801, a noncompetitive NMDA blocker, to inhibit NMDA-elicited cell death. Representative results are shown in Figure 21-3a. In all cases, the added compounds were present from day 0.

a. Which data presented in Figure 21-3a support the hypothesis that activation of NMDA receptors by NMDA causes cell death?

b. Did the embryonic hippocampal cells used in these experiments have NMDA receptors present throughout the two-week study period?

c. Do the data suggest that MK801 might affect the survival of hippocampal neurons by more than one mechanism?

d. As shown in Figure 21-3a, NMDA did not elicit death of hippocampal cells during the first seven days of the study period. How would you distinguish between the two following explanations for this observation: (1) no NMDA receptors are present during this period and (2) receptors are present, but they do not permit entry of Ca^{2+} ions into the cells, which subsequently leads to cell death.

e. From day seven to day 14 of the study period, NMDA caused a substantial increase in cell death. How could you determine if this effect was caused, directly or indirectly, by the influx of Ca^{2+} ions from the extracellular medium into the hippocampal neurons?

f. As shown in Figure 21-3b, the hippocampal neurons exhibit quite different sensitivities to the Ca^{2+} ionophore A23187, which increases the membrane permeability to Ca^{2+} ions, when this drug is added at day two and day 10. Assuming that Ca^{2+} is the major contributor to hippocampal death and NMDA receptors are present throughout the study period, how might the data in Figure 21-3b explain the two-phase + NMDA curve in Figure 21-3a?

g. Isolated hippocampal cells subsequently were incubated with [³H]glutamate for two different time periods after culturing—day seven–14 and day 14–21; the cells were then prepared for quantitative autoradiography. When the fixed cells were examined, the same number of autoradiographic grains were found in the preparations from each culture time-period. Are these results consistent with the data presented in Figures 21-3a and 21-3b?

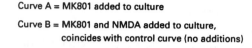

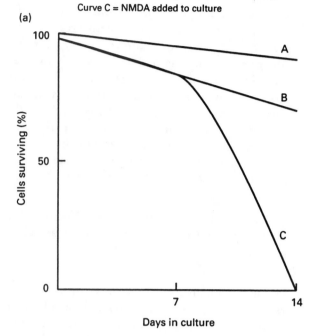

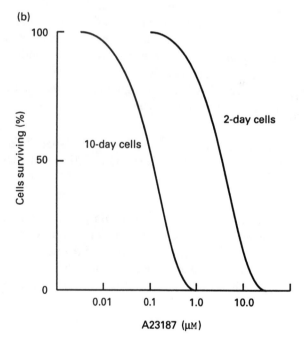

Figure 21-3

Answers

1. A simple reflex arc is the circuit formed by a sensory neuron and a motor neuron. More complicated reflex arcs may involve interneurons, multiple motor neurons, or multiple sensory neurons. An example is the knee-jerk reflex, which involves one sensory neuron, one interneuron, and two motor neurons. The central nervous system includes the brain and the spinal cord and mostly consists of interneurons, while the peripheral nervous system consists of sensory and motor neurons that deliver information to, and receive direction from, the central nervous system.

   For further study, see text section "Neurons Are Organized into Circuits," p. 915, and Figure 21-5, p. 915.

2. Dendrites (usually many per cell) are specialized to receive signals from the axon termini of other neurons or to recognize stimuli as part of sensory detection. Dendrites convert such signals into small electrical disturbances and transmit them toward the cell body. In comparison, the axon (usually one per cell) is specialized for the conduction of action potentials from the cell body to the axon termini (where the signal can be passed on to the next cell in the circuit).

   For further study, see text section "Specialized Regions of Neurons Carry Out Different Functions," p. 912.

3. In nerve-cell signaling, the action potential serves to conduct information (in the form of an electrical disturbance) along the length of the neuron. In comparison, the synapse is the site where the signal (either remaining in electrical form or converted into chemical form) is passed from one cell to another.

   For further study, see text section "Specialized Regions of Neurons Carry Out Different Functions," p. 912; and "Synapses Are Specialized Sites Where Neurons Communicate with Other Cells," p. 914.

4. Sodium channels are inactivated (i.e., closed and insensitive to further voltage changes) for a brief period after channel opening. This property ensures that upstream channels that were previously stimulated will not open again and that downstream channels will be the only ones that can respond to the membrane depolarization.

   For further study, see text section "Action Potentials Are Propagated Unidirectionally without Diminution," p. 923.

5. The myelin sheath acts as an electric insulator for the axon, so that current is not dissipated across the membrane but is channeled to the next node. In addition, ions can flow into a myelinated axon only at the exposed, myelin-free nodes of Ranvier, which contain most of the voltage-gated Na^+ channels in the axon. Thus a region of depolarization (elevated Na^+ levels) can travel along the axon to the next node without having to elicit a response from voltage-gated channels between the nodes. Because the action potential in effect jumps from node to node, the conduction velocity of myelinated axons is greater than that of unmyelinated axons.

   For further study, see text section "Myelination Increases the Velocity of Impulse Conduction," p. 923.

6. Decreasing the external Na^+ concentration, which reduces the magnitude of the Na^+ concentration gradient, therefore causes a decrease in the magnitude of the action potential. Decreasing the external Na^+ concentration to zero would eliminate the action potential altogether, since Na^+ influx could not occur at all.

   For further study see text section "Opening and Closing of Ion Channels Cause Predictable Changes in Membrane Potential," p. 919.

7. Fibroblasts have no voltage-gated ion channels, a necessary prerequisite for the generation and maintenance of action potentials.

For further study, see text section "The Action Potential and Conduction of Electric Impulses," p. 917.

8a. An increase in K⁺ permeability would cause hyperpolarization of the nerve-cell membrane.

8b. A decrease in Cl⁻ permeability would cause depolarization of the nerve-cell membrane.

8c. An increase in Na⁺ permeability would cause depolarization of the nerve-cell membrane.

8d. A decrease in K⁺ permeability would cause depolarization of the nerve-cell membrane.

For further study, see text section "Opening and Closing of Ion Channels Cause Predictable Changes in Membrane Potential," p. 919.

9a. These positively charged amino acids are localized on one side of the conserved α helices and are thought to function as the voltage sensor. When the membrane is depolarized, the sensor helix (S4) is thought to move toward the exoplasmic surface, which is now negatively charged. The conformational change associated with this motion of the sensor helix is thought to induce movement of the channel-blocking segment, thus opening the channel. One S4 helix is present in each of the four transmembrane domains in the Na⁺ and Ca²⁺ channel polypeptides and in each of the four subunits of the K⁺ channel.

9b. Studies with *Drosophila shaker* mutants have confirmed the voltage-sensing role of the S4 helix. The *shaker* gene encodes the K⁺ channel polypeptide; four identical polypeptides associate to form a complete channel in the membrane. Site-directed mutagenesis was used to produce mutant *shaker* genes encoding K⁺ channel proteins in which one or more arginine or lysine residues in the S4 helix was replaced with neutral or negatively charged acidic residues. Oocytes in which these mutant proteins were expressed were shown by patch-clamping studies to exhibit abnormal responses to depolarization voltages: as the number of positively charged residues in the *shaker* K⁺ channel protein was reduced, the depolarizing voltage required to open the channel increased.

For further study, see text section "Voltage-Gated K⁺ Channels Have Four Subunits Each Containing Six Transmembrane α Helices," p. 929; and "The S4 Transmembrane α Helix Acts as a Voltage Sensor," p. 932.

10. A defective Na⁺ channel that remains open and cannot be inactivated in the wild-type manner also could contribute to delayed repolarization of neurons. The best techniques for demonstrating such a defect are voltage clamping and patch clamping.

For further study, see text section "Patch Clamps Permit Measurement of Ion Movements through Single Channels," p. 927.

11. *Xenopus* oocytes do not normally express voltage-gated ion channels, but if injected with mRNA coding for these proteins, the oocytes will express functional channel proteins. These proteins will be inserted correctly in the oocyte plasma membrane and can be studied using patch clamp techniques without any complications due to endogenous voltage-gate ion channels.

For further study, see text section "Voltage-Gated K⁺ Channels Have Four Subunits Each Containing Six Transmembrane α Helices," p. 929.

12. Inhibitory synapses hyperpolarize the postsynaptic cell, whereas excitatory synapses depolarize the postsynaptic cell. An inhibitory response usually involves the opening of K⁺ or Cl⁻ channels. An excitatory response often involves the opening of Na⁺ or, in some cases, the closing of K⁺ channels. Whether an excitatory or inhibitory response is generated at a chemical synapse depends on the neurotransmitter receptor in the postsynaptic-cell membrane. Some neurotransmitters (e.g., acetylcholine) mediate both responses through different receptors, although most are either excitatory (e.g., glutamate and serotonin) or inhibitory (e.g., GABA and glycine).

For further study, see text section "Chemical Synapses Can Be Excitatory or Inhibitory," p. 938.

13. Impulse transmission across electric synapses occurs more rapidly than across chemical synapses.

Thus the two types of synapse can be distinguished by stimulating the presynaptic cell and recording the changes in membrane potential with time in both the pre- and postsynaptic cells. The time course of these changes is characteristic for each type of synapse. In particular, signal transmission across a chemical synapse exhibits a delay of about 0.5 ms. In addition, chemical synapses with specific postsynaptic receptors will respond to specific stimulators and toxins; for example, the nicotinic acetylcholine receptor is stimulated by nicotine and inhibited by α-bungarotoxin. In comparison, transmission across electric synapses is not affected by such modulatory substances.

For further study, see text section "Neurotransmitters, Synapses, and Impulse Transmission," p. 935.

14. In response to membrane depolarization, the voltage-gated Ca^{2+} channels in axon terminals allow an influx of Ca^{2+} from the extracellular space into the presynaptic cell. The Ca^{2+} ions bind to proteins in the membrane of synaptic vesicles that interact with plasma-membrane proteins, inducing membrane fusion and exocytosis of the vesicle contents into the synaptic cleft. Thus these Ca^{2+} channels transduce an electrical signal (membrane depolarization) into a chemical signal (high cytosolic Ca^{2+}), which then induces neurotransmitter release into the synaptic cleft. Since release of neurotransmitter at a chemical synapse depends on the rise in cytosolic Ca^{2+}, incubation of neurons in Ca^{2+}-free medium would prevent impulse transmission from depolarized presynaptic cells to postsynaptic cells.

For further study, see text section "Influx of Ca^{2+} Triggers Release of Neurotransmitters," p. 936.

15. Neurotransmitters are removed from the synaptic cleft by two main mechanisms. Acetylcholine is hydrolyzed to acetate and choline by the extracellular enzyme acetylcholinesterase. For most other neurotransmitters (e.g., dopamine, serotonin, GABA), transport proteins in the plasma membrane of the presynaptic cells retrieve the neurotransmitter molecules to be used again.

For further study, see text section "Transmitter-Mediated Signaling Is Terminated by Several Mechanisms," p. 941.

16. Both Na^+ and K^+ ions are hydrated; that is, a shell of bound water surrounds these cations. The pore of the ion channel in the nicotinic acetylcholine receptor must be large enough to permit passage of hydrated Na^+ and K^+ cations, which are considerably larger than the unhydrated cations.

For further study, see text section "All Five Subunits in the Nicotinic Acetylcholine Receptor Contribute to the Ion Channel," p. 945.

17. Acetylcholine produces opposite effects in skeletal and smooth muscle because the acetylcholine receptors present on the two types of muscle cells are different. In skeletal muscle, the acetylcholine receptor is a ligand-gated cation channels, which when activated by acetylcholine binding, allows Na^+ to enter the cell and depolarize the muscle cell membrane leading to contraction. In smooth muscle, the acetylcholine receptor is a G protein-coupled receptor whose activation leads to opening of K^+ channels, which hyperpolarizes the membrane and slows the rate of contraction.

For further study, see text section "Opening of Acetylcholine-Gated Cation Channels Leads to Muscle Contraction," p. 944; and "Cardiac Muscarinic Acetylcholine Receptors Activate a G Protein That Opens K^+ Channels," p. 948.

18. GABA and glycine are considered inhibitory neurotransmitters because these neurotransmitters activate ligand-gated Cl^- channels. The opening of these channels leads to the slight hyperpolarization of the cell membrane, which in turn briefly makes it more difficult for an action potential to form.

For further study, see text section "GABA- and Glycine-Gated Cl^- Channels Are Found at Many Inhibitory Synapses," p. 947.

19. The sense of taste (salty, sweet, bitter, sour) depends on gated cation channels, which when the stimulus is detected, open to allow a Na^+ influx and subsequent membrane depolarization. In fact, the receptors to detect salt are simply ungated Na^+ channels. In comparison, the detection of odors (the sense of smell) uses G protein-coupled receptors, which activate adenylate cyclase to cause an

increase in the intracellular concentration of the second messenger, cAMP. The cAMP then opens cAMP-gated Na⁺ channels in the olfactory neuron plasma membrane and the membrane is depolarized. There are estimated to be about 1000 different odorant receptors (each specific for a given odorant molecule), but each olfactory neuron expresses only one specific receptor on its surface and, therefore, detects only one or perhaps a few different odorant molecules.

For further study, see text section "Sensory Transduction," p. 951.

20. In the dark, the rod is depolarized and secretes neurotransmitter. When stimulated with light, it hyperpolarizes and secretion of neurotransmitter is reduced. A typical sensory neuron is depolarized when it is stimulated. The rod-cell Na⁺ channel, although similar in sequence to voltage-gated K⁺ channels, contains binding sites for cGMP and opens in response to binding of cGMP. Injection of cGMP into a rod cell leads to opening of these cGMP-gated Na⁺ channels, resulting in further membrane depolarization and enhanced neurotransmitter release at the synaptic end of the rod cell.

For further study, see text section "The Light-Triggered Closing of Na⁺ Channels Hyperpolarizes Rod Cells," p. 952; and "Cyclic GMP Is a Key Transducing Molecule in Rod Cells," p. 954.

21. Aplysia has provided insight into short-term learning processes since it demonstrates the simple forms of learning found in vertebrates: habituation, sensitization, and classical conditioning. In addition, the relatively simple neural network facilitates experimental analysis of the underlying cellular mechanisms responsible for these three processes.

For further study, see text section "Learning and Memory," p. 960.

22a. Generation of an action potential in any postsynaptic cell depends on depolarization of the postsynaptic membrane at the axon hillock to a certain threshold potential. The response of a postsynaptic cell depends on the timing and amplitude of the electric impulses it receives from the presynaptic cells. Figure 21-1b shows that the action potential of neuron C is not in synchrony with those of neurons A and B; thus it does not contribute to generation of the postsynaptic action potential. For this reason, inhibition of neuron C does not prevent firing of the postsynaptic cell. Since inhibition of either neuron A or B prevents firing of the postsynaptic cell, stimulation of both A and B must be required to generate the threshold potential in the postsynaptic cell of this circuit.

22b. One possibility is that the drug maintains the voltage-gated Na⁺ channels in an open position once they have opened, although the drug is not capable of opening voltage-gated channels by itself. Another possibility is that it prevents opening of the voltage-gated K⁺ channels that are responsible for the quick return to the resting potential.

22c. According to Figure 21-3, the neurotoxin decreases both the resting potential and the action potential of the neuron. Increasing the external K⁺ concentration would cause a decrease in the resting potential, and decreasing the external Na⁺ concentration would cause a decrease in the action potential.

For further study, see text section "The Action Potential and Conduction of Electric Impulses," p. 917.

23a. These data indicate that free synapsin I can bind to stripped synaptic vesicles with half-maximal binding occurring at a synapsin I concentration of about 50 nM in 40 mM NaCl. High salt concentrations, however, reduce the binding of synapsin I to the vesicles.

23b. The reduction in synapsin I binding following protease treatment of vesicles suggests that the stripped vesicles contain a receptor protein on their surface that facilitates binding of synapsin I. The reduced binding at 150 mM NaCl suggests that electrostatic interactions are important in the binding of synapsin I to this receptor protein.

23c. These results suggest that the tail domain of synapsin I binds to a protein receptor on the vesicle but that the head merely interacts nonspecifically with phospholipids in the vesicle membrane.

23d. This experiment suggests that added synapsin I probably is incorporated into stripped vesicles in a manner similar to that which occurs *in vivo*.

23e. The most definitive experiment would be to subject the isolated proteins to Western-blot analysis using antibodies against synapsin I.

For further study, see text section "Multiple Proteins Participate in Docking and Fusion of Synaptic Vesicles," p. 936.

24a. The hypothesis is supported by the observation that cell death is much greater in the presence than in the absence of NMDA (+NMDA curve versus control curve). In addition, the specificity of this effect is indicated by the ability of MK801, a blocker of NMDA, to inhibit it; that is, the survival curve in the presence of MK801 + NMDA is similar to the control curve.

24b. The absence of any effect of NMDA on cell survival during the first seven days in culture suggests that either no NMDA receptors or no functional receptors were present in the embryonic hippocampal cells during this period. Additional studies would be needed to confirm this conclusion, however (see part d).

24c. MK801 alone increased the survival rate from day one in culture compared with control cultures, whereas MK801 + NMDA blocked the NMDA-elicited rapid cell death beginning at day seven. These findings suggest that MK801 can inhibit cell death by two different mechanisms, one of which prevents activation of NMDA receptors by NMDA and the other of which does not involve these receptors.

24d. The best way to distinguish these explanations would be binding studies with [³H]glutamate, using NMDA to displace the radioactive glutamate. If receptors are present, then binding should be observed; in this case, presumably, the receptors, though present, do not allow an influx of Ca^{2+} ions early in the culture period.

24e. One way to demonstrate whether NMDA-elicited cell death is related to the influx of Ca^{2+} ions would be to compare the NMDA effect at high and low external Ca^{2+} concentrations. One problem with this experimental approach is that cells need some extracellular Ca^{2+} to remain attached to the substratum; thus depleting external Ca^{2+} might cause some effects unrelated to the NMDA effect.

24f. The data in Figure 21-3b indicate that an approximately 100-fold higher concentration of Ca^{2+} ionophore is required to cause 50-percent cell death in 2-day cells than in 10-day cells, presumably because the younger cells are less permeable to Ca^{2+}. These findings suggest that the inability of NMDA to elicit cell death during the first seven days of culture, as shown in Figure 21-3a, results from the inability of the stimulated receptor to permit an influx of Ca^{2+} ions from the extracellular medium.

24g. This experiment provides no useful information because the glutamate binding observed may reflect nonspecific binding to various cell structures and/or binding to the other two glutamate receptor subtypes. In order to demonstrate the presence of the NMDA receptor, competition experiments using NMDA to displace the radioactive glutamate are necessary.

For further study, see text section "Two Types of Glutamate-Gated Cation Channels May Function in a Type of Cellular Memory," p. 946.

22

Integrating Cells into Tissues

PART A: *Chapter Summary*

The ability of cells to adhere tightly and interact specifically with each other is a key step in the evolution of multicellularity. Cell-adhesion molecules (CAMs) enable many animal cells to adhere tightly and specifically with cells of the same, or similar, type. These interactions allow populations of cells to segregate into distinct tissues. These interactions are stabilized by elaborate specialized cell junctions which also promote local communication between adjacent cells. Animal cells secrete a complex network of proteins and carbohydrates, the extracellular matrix (ECM), that creates a special environment in the spaces between cells. The matrix helps bind the cells in tissues together and is a reservoir for many hormones that control cell growth and differentiation. The matrix also provides a lattice through which cells can move, particularly during the early stages of differentiation. Defects in cell-cell and cell-matrix connections lead to cancer and developmental malformations.

The extracellular matrix in animals has three major protein components: highly viscous proteoglycans, which cushion cells; insoluble collagen fibers, which provide strength and resilence; and soluble multi-adhesive matrix proteins, which bind these components to receptors on the cell surface. Different combinations of these components tailor the strength of the extracellular matrix for different purposes. In this chapter, we focus on the role of structure and of interactions between ECM components, cell-adhesion molecules, and cell-adhesion in permitting animal cells to form organized tissues. Plant cells are surrounded by a cell wall that is thicker and more rigid than the animal cell extracellular matrix. Although the plant cell wall and the extracellular matrix serve many of the same functions, they are structurally very different. Because of these differences, the plant cell wall and its interactions with plant cells is presented in a separate section at the end of the chapter.

PART B: *Reviewing Concepts*

22.1 *Cell-Cell Adhesion and Communication* (lecture date _____)

1. Cadherins are known to mediate homophilic interactions between cells. What is a homophilic interaction and how can this be demonstrated experimentally for E- and P-cadherins?

2. What is the role of selectins in leukocyte extravasation?

3. How do cadherins in adherens junctions and spot desmosomes intreact with cytoskeletal proteins?

22.2 *Cell-Matrix Adhesion* (lecture date _____)

4. To what extent are integrins encoded by a single gene versus multiple genes?

5. Cell-matrix interactions through integrins hold cells in place. How can such associations be modulated to promote cell motility?

6. What distinguishes a hemidesmosome from a focal adhesion?

22.3 *Collagen: The Fibrous Proteins of the Matrix* (lecture date _____)

7. Heating of calf type I collagen fibers to 45°C denatures the triple helices and separates the three chains from each other. Collagen that has been treated in this way does not renature to form a normal collagen triple helix. Why?

8. Even low levels of expression of mutant α1(I) collagen, 10% of wild type, cause severe growth abnormalities in transgenic mice. Why does even low-level expression suffice to cause abnormalities?

9. Type IX collagen does not form fibrils, but does associate with collagen fibrils composed of type II collagen. What structural characteristics of type IX collagen are responsible for this functional distinction?

22.4 *Noncollagen Components of the Extracellular Matrix* (lecture date _____)

10. What are the two major classes of soluble extracellular matrix proteins in vertebrates?

11. Extracellular matrix components in animals are often mostly carbohydrate, and, in fact, hyaluronan contains no protein component. How do the saccharide portions of these molecules contribute to their biological role?

12. How does the structure of fibronectins support their role as a multi-adhesive matrix proteins?

22.5 *The Dynamic Plant Cell Wall* (lecture date _____)

13. What properties of plant cells are thought to be responsible for the fact that all plant hormones are small and water soluble?

14. What is the role of expansin and pH in auxin-induced plant cell growth?

15. Plant cell-wall molecules have many functional, but not necessarily structural, homologies with molecules found in the extracellular matrices around animal cells. Three important plant cell-wall constituents are (a) cellulose, (b) pectin, and (c) lignin. Describe the functions performed by each of these plant cell wall compounds.

PART C: *Analyzing Experiments*

22.1 *Cell-Cell Adhesion and Communica-tion*

16. (•••) Gap junctions, that form between adjacent cells in an epithelium, mediate exchange of small molecules and ions between the cells. This exchange may be important in many cellular regulatory processes, including those that regulate cell growth. Some researchers have conducted experiments to determine whether cancerous cells exhibit junctional communication.

In one experiment, depicted in Figure 22-1a, four normal cells were surrounded by cancerous cells. A small amount of fluorescent dye (MW = 310) was injected into one of the normal cells. After a few minutes, the cells were examined, using a fluorescence microscope to determine whether the dye had spread to other cells. Fluorescent cells are indicated by crosshatching.

In a second experiment, depicted in Figure 24-1b, a small amount of the fluorescent dye was injected into one of the cancerous cells surrounding the four normal cells. Again, the cells were examined under a fluorescence microscope after a few minutes.

a. Are these data consistent with the hypothesis that normal cells are coupled to each other via gap junctions? Why or why not?

b. Are these data consistent with the hypothesis that cancerous cells are coupled to each other via gap junctions? Why or why not?

c. Are these data consistent with the hypothesis that cancerous cells and normal cells are coupled to each other via gap junctions? Why or why not?

d. Describe an experimental approach that could be used to determine whether the cellular communication demonstrated with this fluorescence assay occurs via gap junctions.

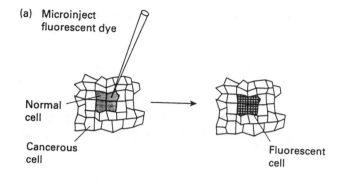

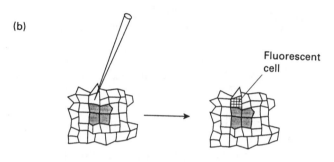

Figure 22-1

22.1 *Cell-Cell Adhesion and Communication, 22.5 The Dynamic Plant Cell Wall*

17. (•) In order to prepare single-cell suspensions of most epithelial cells grown in culture, scientists add calcium-chelating reagents such as EDTA or EGTA to confluent monolayers of these cells. In many cases trypsin must be added as well.

    a. What is the probable action of these agents in dissociating epithelial cells?

    b. What enzymes would you choose if you wanted to dissociate cultured plant cells?

22.1 *Cell-Cell Adhesion and Communication, 22.2 Cell-Matrix Adhesion*

18. (•) Although there are many different types of transmembrane adhesion molecules, they do share some common functions. These molecules (e.g., cadherins, integrins, N-CAMs, integral-membrane proteoglycans) have relatively low affinities for their respective ligands, but are found at relatively high concentrations on cell surfaces. Additionally, many are linked to cytoskeletal networks on the cytosolic surface of the plasma membrane. What functions are served by these properties of cell-surface adhesion molecules?

22.3 *Collagen: The Fibrous Proteins of the Matrix*

19. (•) Some of the symptoms of vitamin C deficiency in humans include loss of teeth, lesions on the skin, and weakening of the blood vessels. From what you know of the involvement of vitamin C in collagen biochemistry, synthesis and intracellular processing, what may be concluded from these observations about the metabolic stability of collagen in gum, skin, and vascular tissues in adult human beings? Explain your answer.

20. Many connective tissue diseases result from synthesis of aberrant, but secreted collagens. Some of these collagens can be distinguished from native collagens by their different migratory pattern on a denaturing (SDS) gel. SDS gel electrophoresis of collagen from a child who exhibits a possible collagen-deficiency disease produced a profile indicative of a type I collagen deficiency. When the separated chains were extracted from the gel and recombined in an appropriate buffer, they failed to reassociate into a native triple helix. Do these data indicate that the molecular defect in this child involves the inability of collagen to form its triple helix once secreted into the extracellular matrix?

22.5 *The Dynamic Plant Cell Wall*

21. (••) Expansins are cell wall proteins involved the acid pH dependent stretching of the plant cell well. As shown in Figure 22-2, both the enzyme cellulase and expansin cause cell wall stretching in an elastometer assay.

    a. What is the importance of the boiled expansin control, pH 4.5?

    b. Propose an experimental test of whether either cellulase or expansin produces a reversible stretching of the cell wall.

    c. Assuming that cellulase produces a non-reversible stretching while expansin produces a reversible stretching, what interpretation may be made as to whether or not expansin cleaves covalent bonds?

(a)

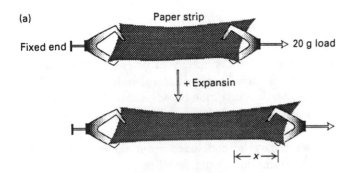

(b)

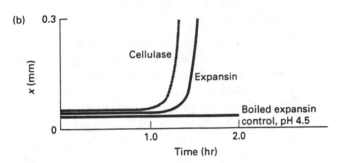

Figure 22-2

Answers

1. Homophilic interactions are those between like cell types, e.g., epithelial cells with epithelial cells. One approach to demonstrating homophilic cell interactions experimentally is to use L cell lines transfected with either E-cadherin or P-cadherin. L cells adhere poorly to each other and express no cadherins. When transfected with E- and P-cadherin, L cells adhere tightly to E-cadherin-postive cell to E-cadherin-positive cell and P-cadherin-positive cell to P-cadherin-positive cell. Cadherins directly cause homotypic interactions among cells.

   For further study, see text section "Cadherins Mediate Ca^{2+}-Dependent Homophilic Cell-Cell Adhesion." p. 971.

2. Extravasation is the movement into tissues of four types of leukocytes (white blood cells). This is particularly important for monocytes, the precursors of macrophages, which ingest foreign particulates; neutrophils, which release several antibacterial proteins; and T and B lymphocytes, the antigen-specific cells of the immune system. These cells are present in the blood stream and carry on their surfaces selectin-specific ligands (the sialyl Lewis-x antigen). Endothelial cells facing onto the blood

vessel become activated in response to various inflammatory signals released by surrounding cells in areas of infection or inflammation. Activated endothelial cells exocy-tose P-selectin to their cell surface. As a consequence, passing leukocytes adhere weakly to the endothelium; because of the force of the blood flow, these trapped leukocytes are slowed but not stopped and seem to roll along the surface of the endothelium.

In later steps, binding of leukocytes to the endothelium becomes tight through integrins and the leukocyte inserts itself between endothelial cells.

For further study, see text section "Selectins and Other CAMs Participate in Leukocyte Extravasation," p. 972.

3. Cadherins bridge between neighboring cells in adherens junctions and desmosomes. The cadherins link to adapter proteins, which then link to cytoskeletal proteins. In adherens junctions, the link is to F-actin and myosin in the circumferential belt located below the tight junction of the epithelial cell layer and in desmosomes, to the intermediate filament protein keratin. In the case of desmosomes the adapter complex appears distinguishable as a cytoplasmic plaque. The plaque proteins are plakoglobin and desmoplakins. In adherens junctions, adapter proteins include α- and β-catenin.

   For further study, see text section "Caherin-Containing Junctions Connect Cells to One Another," p. 973.

4. Mammals have at least 22 different integrin heterodimers. These are composed of differing α and β subunits. Multiple α and β subunits are coded for by separate genes rather than being different splicing forms. Because these are heterodimers and hence combination products, the actual number of genes needed is smaller than the number of different heterodimers.

   For further study, see text section "Integrins Mediate Weak Cell-Matrix and Cell-Cell Interactions," p. 977.

5. Various de-adhesion factors promote cell migration. One class of de-adhesion factors comprises small peptides called disintegrins. They contain

the inegrin-binding RGD sequence present in many ECM proteins. By binding to integrins on the surface of cells, disintegrins competitively inhibit binding of cells to matrix components. A second class of de-adhesion factors includes two types of proteases, fibrinogens and matrix-specific metalloproteinases. Both proteases degrade matrix components, thereby permitting cell migration.

Other de-adhesion factors such as ADAMs containing *a* *d*isintegrin and *a* *m*etalloprotease participate in events that depend on the remodeling of the cell surface. These are not primarily cell migration related.

For further study, see text section "De-adhesion Fators Promote Cell Migration and Can Remodel the Cell Surface," p. 978.

6. Focal adhesions and hemidesmosomes are cell-matrix junctions that consist of clustered integrin molecules. The extracellular domain of integrins in a focal adhesion binds to fibronectin, and in hemidesmosomes, to laminin in the basal lamina. The intracellular domain binds to adapter proteins. In focal adhesions the adapter proteins link to actin-rich stress fibers. In hemidesmosomes, the adapter proteins link to keratin containing intermediate filaments.

For further study, see text section "Integrin-Containing Junctions Connect Cells to the Substratum," p. 978.

7. The N-terminal and C-terminal propeptides present in newly synthesized collagen monomers assist in alignment of the peptides to form the triple helix. These propeptides are removed after the trimers are transported to the extracellular matrix, and thus are not available to perform the same function in denatured calf type I collagen. In addition, inappropriate disulfide bridges can be generated during renaturation; these will also inhibit the generation of a normal triple helix.

For further study, see text section "Assembly Collagen Fibers Begins in the ER and Is Completed outside the Cell," p. 981.

8. Each type I collagen molecule contains two $\alpha 1(I)$ and one $\alpha 2(I)$ chains. Hence expression at low frequency for the $\alpha 1(I)$ chain will have more effect

than the same frequency of expression of an $\alpha 2(I)$ chain mutation. The polymerization of the triple helix and its formation into fibrils is affected.

For further study, see text section "Mutations in Collagen Reveal Aspects of Its Structure and Biosynthesis," p. 982.

9. Type IX collagen consists of two triple-helical domains separated by a flexible kink. A proteoglycan is attached at and protrudes from this kink region. The interruption in the triple helix, as well as the presence of the proteoglycan, prevents this molecule from self-associating to form collagen fibrils. Type IX collagen associates with other type II collagen fibrils but cannot form fibrils itself.

For further study, see text section "Collagens Form Diverse Structures," p. 984.

10. The two major classes of soluble extracellular matrix proteins are multi-adhesive matrix proteins, which bind cell-surface adhesion receptors, and proteoglycans, a diverse group of macromolecules containing a core protein with multiple attached polysaccharide chains.

For further study, see text sections "Noncollagen Components of the Extracellular Matrix," p. 985 and "Hyaluronan Resists Compression and Facilitates Cell Migration," p. 992.

11. Because of their high content of charged polysaccharides, proteoglycans and free hyaluronan, which is entirely carbohydrate, are highly hydrated. The swelled, hydrated structure of proteoglycans is largely responsible for the volume of the extracellular matrix and also acts to permit diffusion of small molecules between cells and tissues. The swelling forces of water associated with hyaluronan, for example, give connective tissues their ability to resist compression forces. The loose, hydrated nature of hyaluronan is particularly important in permitting cell migration.

For further study, see text section "Noncollagen Components of the Extracellular Matrix," pp. 985–992.

12. Fibronectins are an important class of soluble multi-adhesive matrix proteins. The multi-adhesive properties of fibronectins arise from the

presence on different domains of high-affinity binding sites for collagen and other ECM components and for certain integrins on the surface of cells. Hence, the structure leads to the ability to bind to multiple matrix components and to cell surface proteins.

For further study, see text section "Fibronectins Bind Many Cells to Fibrous Collagens and Other Matrix Components," p. 987.

13. The porosity of the plant-cell wall permits soluble factors to diffuse and interact with receptors on the plant plasma membrane. Water and ions diffuse freely in cell walls but diffusion of even relatively small proteins is restricted. This then is one of the reasons that plant hormones are small, water-soluble molecules.

For further study, see text section "The Dynamic Plant Cell Wall," p. 993.

14. Auxin stimulates proton secretion at the growing end of the cell. The low pH activates a class of cell wall proteins, termed expansins. Expansins disrupt the hydrogen bonding between cellulose fibrils causing the laminate structure of the cell wall to loosen. With the rigidity of the wall reduced, the cell can elongate.

For further study, see text section "A Plant Hormone, Auxin, Signals Cell Expansion," p. 996.

15. (a) Cellulose is analogous to collagen. Both are long, insoluble, fibrous polymers, and in both cases the fibers are generated extracellularly. Both confer tensile strength to their respective tissues, even though one is a carbohydrate and one is a protein. (b) Pectin is analogous to hyaluroran. Both are multiple negatively charged polysaccharides, with a high binding capacity for water. Both bind to other matrix molecules and allow tissues to resist compressive forces. (c) Lignins are analogous to proteoglycans. Both bind to other components of the matrix, enabling tissues to resist compressive forces.

For further study, see text sections "The Cell Wall Is a Laminate of Cellulose Fibrils in a Pectin and Hemicellulose Matrix," p. 993; and "Cell Walls

Contain Lignin and an Extended Hydroxyproline-Rich Glycoprotein," p. 995.

16a. Yes. The fluorescent dye spreads to adjacent cells; this is consistent with the known properties of gap junctions.

16b. No. The dye is confined to the injected cell. If the cells contained functional gap junctions, the dye should spread to adjacent cells.

16c. No. Dye injected into a normal cell does not spread to adjacent cancerous cells. Dye injected into a cancerous cell does not spread to adjacent normal cells.

16d. The simplest approach involves injecting a dye with a molecular weight greater than 2000. A dye this large should not transfer to adjacent normal cells via gap junctions, which only permit ready passage of quite small molecules. Molecules with a molecular weight of about 1200 pass easily, whereas those with molecular weights above 2000 are excluded. Passage of intermediate-sized molecules across gap junctions is limited and variable. Other approaches such as inhibitory antibodies against connexin could also be used.

17a. Calcium chelators remove calcium from E-cadherin, thus facilitating dissociation. Trypsin degrades cadherin and other cell-surface adhesion molecules, thus preventing reassociation.

17b. Adhesion of plant cells is affected by interactions between components of the cell wall around the cells. Because cellulose and pectin are major components of the cell wall, pectinase and cellulase would be the most appropriate enzymes for dissociating plant cells.

18. Low-affinity receptors must be present at high concentrations in order to affect binding of cells to each other or to the extracellular matrix. This is analogous to Velcro™, where each individual link does not have a high bond strength, but the sum total of all the linkages is quite strong.

The linkage of cell surface adhesion molecules to cytoskeletal networks may serve two functions. This association primarily serves to further strengthen cell-cell adhesion against stresses from stretching or pulling of the matrix. If the transmembrane adhesion molecules were not attached to something inside the cell, they could easily be pulled from the membrane by forces that are routinely encountered during epithelial folding and stretching. Again, a mechanical analogy might be a *molly* (Figure 22-3). This device is a bolt that passes through a thin wall or board and is attached to a spring-loaded clip on the far side of the board. Lateral expansion of the clip prevents the bolt from being pulled out of the hole through which it passed originally. Additionally, the association of cell-surface adhesion molecules with cytoskeletal filaments may serve to assist and stabilize the lateral clustering of adhesion molecules, which is necessary for the development of multiple small binding sites (focal adhesion plaques).

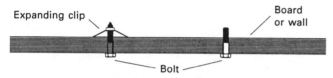

It would take significantly more energy to remove the left bolt than the right bolt from the board because the attached expanding clip would also have to be pulled through the board material.

Figure 22-3

19. These observations indicate that matrix collagen is dynamic, rather than static: Vitamin C is a cofactor for the hydroxylation of proline residues in collagen in the ER; these hydroxylated proline residues are important for the stability of the collagen triple helix. Nonhydroxylated procollagen chains, which would be found in cases of vitamin C deficiency, are degraded within the cell and never secreted to form collagen fibrils. The finding that gum tissue, skin tissue, and vascular tissue are all weakened in cases of vitamin C deficiency indicates that normally the collagen fibers in these tissues must be degraded and replaced regularly. Loss of collagen fibers without subsequent replacement, as occurs in cases of vitamin C deficiency, leads to weakening of the tissues.

20. No. Although some collagen diseases are thought to result from an inability to assemble triple-helical collagen in the extracellular matrix, all collagens need N- and C-terminal propeptides to aid in the formation of the triple helix. These propeptides are cleaved and not present in the mature, extracellular matrix collagen. Thus even denatured type I collagen from normal individuals would not renature to form the native triple helix under the experimental conditions described.

21a. The boiled expansin control demonstrates that the expansin must be in its native configuration to work and that acid pH alone does not suffice. Boiling denatures proteins.

21b. The stretched paper could be washed with water or a neutral salt buffer solution to remove cellulase or expansin. The paper could then be incubated a neutral pH and scored for whether or not it contracts back to much of its former size. If contraction is observed, then the expansion is reversible.

21c. Reversible stretching of the washed paper would indicate that expansin did not cleave covalent bonds. Cleavage of covalent bonds is an irreversible effect.

Cell Interactions in Development

PART A: *Chapter Summary*

Local interactions between cells, mediated by secreted or cell-surface signaling molecules, determine regionalization along the dorsoventral axis in *Drosophila* and along both major axes in early vertebrate embryos. Such local interactions also are the primary mechanism regulating the formation of internal organs such as the kidney, lung, and pancreas. The importance of cellular interactions in development was demonstrated first in the early part of the twentieth century in a series of experiments establishing the principle of induction. Induction refers to any mechanism whereby one cell population influences the development of neighboring cells. In some cases induction involves a binary choice. In other cases, signals can induce different responses in cells at different concentrations. The concentration at which a signal induces a specific cellular response is called a threshold. In many cases, an inductive signal induces an entire tissue containing multiple cell types. The two models proposed to account for these properties of extracellular signaling are the gradient model, in which a signaling molecule induces different fates at different threshold concentrations and the relay model, in which a signal induces a cascade of induction in which cells close to the signal source are induced to assume specific fates and they, in turn, produce other inductive signals. Inductive interactions are often unidirectional and are sometimes reciprocal. Inductive interactions may occur between equivalent cells or non-equivalent cells. Cells differ greatly in their ability to respond to inductive signals. Cells that can respond to such signals are referred to as competent.

In this chapter, we first describe examples of various types of inductive signals and cellular interactions that regulate cell-type specification in several different developmental systems. Specific extracellular signals also control the migration of certain cells, which occurs during development of some tissues. As an example of this phenomenon, the role of extracellular signals in the assembly of connections between neurons is discussed. Another common feature of developmental programs is the highly regulated death of certain cells. The final section of this chapter examines the conserved pathway leading to cell death and how it is controlled. The examples presented are chosen to illustrate key concepts.

PART B: *Reviewing Concepts*

23.1 *Dorsoventral Patterning by TGFβ-Superfamily Proteins* (lecture date _____)

1. TGFβ receptors fall into three classes, type I, II, and III. What is the chemical nature of each and what is their individual contribution to the overall functioning of TGFβ as an inducer.

2. Discuss how microinjection experiments using mRNA encoding Dpp, a *Drosophila* TGFβ superfamily protein, provide evidence for the importance

of inducer thresholds in controlling dorsoventral cell patterning.

3. What is Spemann's organizer?

23.2 *Tissue Patterning by Hedgehog and Wingless* (lecture date _____)

4. Compared to wingless, hedgehog is a non-diffusible inductive signal. How is the diffusion of hedgehog limited?

5. The zone of polarizing activity (ZPA) in the chick limb bud induces wing formation. What is the result of transplanting ZPA to the anterior region of the normal limb bud (ectopic expression)?

23.3 *Molecular Mechanisms of Responses to Morphogens* (lecture date _____)

6. Define the term morphogen.

7. Does differentiation of cells appear to depend on the absolute or relative number of receptor molecules occupied? Give an example.

23.4 *Reciprocal and Lateral Inductive Interactions* (lecture date _____)

8. Salivary-gland differentiation requires interactions between mesenchyme and endodermal cells. What is the role of the basal lamina in these interactions?

9. In lateral interactions between equivalent cells in *C. elegans* to give rise to AC and VU, is there a predetermined pattern for which of the two cells becomes which?

23.5 *Overview of Neuronal Outgrowth* (lecture date _____)

10. Is the end distribution of the axon of a neuron controlled by the molecules expressed by the neuron alone, those expressed by the other cells or by a combination of the two?

23.6 *Directional Control of Neuronal Outgrowth* (lecture date _____)

11. Describe an experimental demonstration using the floor plate as an example that chemoattractants play a role in directing neuronal outgrowth in vertebrates.

12. *Unc-5, unc-6,* and *unc-40* are genes involved in the dorsoventral outgrowth of neurons in *C. elegans*. These genes were identified by mutations. The vertebrate equivalent of UNC-6 is neutrin. Which of the gene product(s) are receptors and which are ligand(s)?

23.7 *Formation of Topographic Maps and Synapses* (lecture date _____)

13. What is a topographic map?

14. Ephrin A2 is a growth cone repellent. How does a repellent produce differential formation of neural connections?

23.8 *Cell Death and Its Regulation* (lecture date _____)

15. What is apoptosis?

16. Within the sequence of events leading to programmed cell death, where do caspases act?

PART C: *Analyzing Experiments*

23.2 *Tissue Patterning by Hedgehog and Wingless*

17. (••) Wingless is a diffusible signaling molecule in *Drosophila*. Wingless is produced by a single strip of cells within *Drosophila* body segments (darkened cells, Figure 23-1, A = anterior, P = posterior). Wingless activity results in the specification of a naked cuticle indicated in cartoon form in Figure 23-1 by the absence of bristles above the cells. In the presence of hedgehog expression, wingless protein is found asymmetrically distributed across the body segment as indicated by the arrows. Correspondingly the development of naked cuticle is asymmetric. In the absence of Hedgehog expression, the distribution of Wingless is symmetric as is the development of naked cuticle.

+ Hedgehog expression

- Hedgehog expression

Figure 23-1

a. Describe two ways by which an asymmetric distribution of wingless could arise.

b. Is hedgehog expressed by a cell anterior (A) or posterior (P) to the cell expressing wingless?

c. Is the Hedgehog expressing cell distal or adjacent to the Wingless expressing cell?

23.5 *Overview of Neuronal Outgrowth*

18. (••) Neurons from mice grown in culture provide a useful experimental system. These cells in culture extend processes, i.e., neurites, and one of these neurites acts dominant and becomes the axon. The axon is distinguished visually by its length. It is much longer than the other processes (Figure 23-1). Neurons form one and only one axon. The other neurites become the dendrites of the neuron.

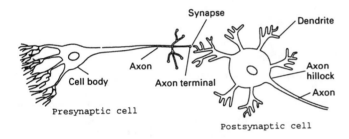

Figure 23-1

Investigators have asked the question of whether any neurite is capable of becoming an axon. To do this, early neurons having neurites but not yet axons were tested. In the experiment, micropipettes were used to locally apply at low concentrations the drug cytochalasin D (CD) at individual neurite endings. CD disrupts actin-containing microfilaments. What was observed was that for any neuron, CD treatment of an individual neurite resulted in that neurite becoming the axon. That neurite began to grow out rapidly.

a. Do these results suggest that a stable actin meshwork in the neurite growth cone is a positive or a negative factor in controlling the transition from neurite to axon?

b. Do these results suggest that the neurite to axon transition is predetermined or a random event?

c. What evidence does this experiment provide regarding the question of why cells have only one axon and not several?

23.8 *Cell Death and Its Regulation*

19. (•••) DCP-1 is a *Drosophila* caspase. *Drosphila* express other proteins which are thought to interact with DCP-1. These include DIAP1, Head Involution Defective (HID) and GRIM. Figure 23-3A shows the effect of expressing either DCP-1 alone or DCP-1 in the presence of DIAP1 alone or plus HID or GRIM. DCP-1 expression is under the control of a GAL inducible promoter. In glucose grown yeast, DCP-1 is not expressed. The darkened areas in Figure 23-3A are yeast colonies formed over a series of 10-fold dilutions in cell number.

In Figure 23-3B, the results show an affinity column binding assay for physical interactions between DCP-1, HID, and HIDΔ(1-36) protein with column bound GST-DIAP1 chimeric protein. The first three proteins were all purified as His6 tagged proteins. HIDΔ(1-36) has a deletion of amino acids 1-36. The results shown are a Western blot to identify possible column binding proteins using specific antibodies. The column matrices are either GST alone or the GST-DIAP1 chimeric protein.

a. What is the effect of DCP-1 expression on yeast colony formation?

b. What is the effect of coexpression of DCP-1 and DIAP1 on yeast colony formation?

c. What is the effect of HID or GRIM on yeast colony formation during coexpression of DCP-1 and DIAP1?

d. Which molecules are apparent inducers of DCP-1 caspase activity and which are inhibitors?

e. What proteins bind to DIAP1?

f. Where within HID is the DIAP1 binding domain?

A

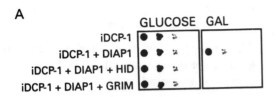

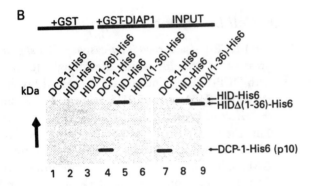

Figure 23-3

Answers

1. Chemical crosslinking experiments indicate TGFβ binds to three different types of receptors—type I, II, and III. Type I and II are both transmembrane proteins with serine/threonine protein kinase activity. The type III receptor is a cell-surface proteoglycan called β-glycan. Although a protein, it is very rich in carbohydrate. It appears to present TGFβ to type I and II receptors.

   For further study, see text section "TGFβ Proteins Bind to Receptos That Have Serine/Threonine Kinase Activity," p. 1005.

2. Microinjection experiments demonstrate that different levels of Dpp specify different cell fates along the dorsoventral axis. This suggests that Dpp acts as a morphogen to induce establishment of different epidermal cell types at different threshold concentrations.

   For further study, see text section "Dpp Protein, a TGFβ Homolog, Controls Dorsoventral Patterning in *Drosophila* Embryos," p. 1007.

3. Spemann's organizer is a region of the dorsal mesoderm that specifies the organization of the mesoderm along the dorsoventral axis through the production of a horizontal dorsalizing signal.

For further study, see text section "Sequential Inductive Events Regulate Early *Xenopus* Development," p. 1007.

4. Hedgehog is tethered to the cell surface by the attachment of cholesterol to the C-terminus of the 20-kDa N-terminal fragment of hedgehog precursor.

   For further study, see text section "Modification of Secreted Hedgehog Precursor Yields a Cell-Tethered Inductive Signal," p. 1013.

5. Transplanting the ZPA to the anterior region of the chick wing bud induces ectopic expression of a second wing.

   For further study, see text section "Hedgehog Organizes Pattern in the Chick Limb and *Drosophila* Wing," p. 1014.

6. Morphogens are inductive signals that lead to qualitatively different developmental responses depending on their concentration.

   For further study, see text section "Molecular Mechanisms of Responses to Morphogens," p. 1018 and in particular "Summary: Molecular Mechanisms of Responses to Morphogens," p. 1021.

7. The dependence appears to be on the absolute number of receptor molecules occupied not the relative number. This was demonstrated for the induction of *Xenopus brachyury* gene.

   For further study, see text section "Cells Can Detect the Number of Ligand-Occupied Receptors," p. 1019.

8. The differentiation of the salivary gland requires the presence of basal lamina. The basal lamina must be between the endodermally derived cells and the mesenchyme cells. It must be there but, for now, its exact role is not known.

   For further study, see text section "The Basal Lamina Is Essential for Differentiation of Many Epithelial Cells," p. 1024.

9. The end outcome of the two equivalent cells to produced AC and VU cells is not set. Either cell could have either fate.

For further study, see text section "Interactions between Two Equivalent Cells Give Rise to AC and VU cells in *C. elegans*," p. 1026.

10. The end distribution of an axon of a neuron is the result of a combination of complex interactions between neuron expressed molecules and molecules expressed by other cells.

For further study, see text section "Soluble Graded Signals Can Attract and Repel Growth Cones," p. 1034 and the text section "Overview of Neuronal Outgrowth," on p. 1028.

11. Explants of vertebrate dorsal spinal cord and floor plate can be arranged in a collagen gel in such a manner to provide an experimental test of whether the floor plate secretes a neural attractant. The secretion of an attractant will cause the nerve endings to grow towards the floor plate.

For further study, see text section "Vertebrate Homologs of *C. elegans* UNC-6 Both Attract and Repel Growth Cones," p. 1034.

12. UNC-5 and UNC-40 are cell surface receptors. UNC-6 (netrin) is a diffusible protein ligand that binds to UNC-5 and UNC-40 as receptor proteins.

For further study, see text sections "Vertebrate Homologs of *C. elegans* UNC-6 Both Attract and Repel Growth Cones," p. 1034; "UNC-40 Mediates Chemoattraction in Response to Netrin in Vertebrates," p. 1036; and "UNC-5 and UNC-40 Together Mediate Chemorepulsion in Response to Netrin," p. 1036.

13. Topographic maps are the spatial description of the relationship between neurons and their targets. They can be thought of as the equivalent of wiring diagrams.

For further study, see text section "Formation of Topographic Maps and Synapses," p. 1038.

14. The EphA3 receptor binds the repellent ephrin A2. Cells expressing high amounts of the receptor will be repelled at low concentrations of ephrin A2, i.e., early, while those expressing low amounts of

receptor will be sensitve to only high levels of the repellent and hence repelled late.

For further study, see text section "The EphA3 Receptor Is Expressed in a Nasal-Temporal Gradient in the Retina," p. 1041.

15. Apoptosis is defined as the demise of a cell through a well-defined sequence of morphological changes constituting programmed cell death.

For further study, see text sections "Programmed Cell Death Occurs through Apoptosis," p. 1045.

16. Caspases, as illustrated in MCB, 4e, Figures 23-49 and 23-50 act relatively late in the cascade of events that produce apoptotic cell death.

For further study, see text sections "Pro-Apoptotic Regulators Promote Caspase Activation," and "Some Trophic Factors Prevent Apoptosis by Inducing Inactivation of a Pro-Apoptotic Regulator" p. 1048.

17a. An asymmetric distribution of wingless could arise from either the existence of a barrier to diffusion of wingless or to the localized degradation of wingless.

17b & c. Mature hedgehog is tethered through conjugation with cholesterol on the cell surface. Hence it should be expressed by the cell immediately posterior to the cell expressing wingless. It has to be the adjacent cell because hedgehog can not diffuse. De-expression of hedgehog changes wingless distribution only in the posterior direction.

18a. The results suggest that a stable actin network in the growth cone is inhibitory, i.e., a negative factor, in controlling axon formation. Cytochalasin D (CD) disruption of actin meshworks promotes axon outgrowth.

18b. Since any neurite can apparently become an axon, the transition must be a random event.

18c. This experiment provides no evidence as to why only one axon forms. It suggests that once a neurite starts to elongate rapidly that it somehow inhibits the other neurites from doing the same.

19a. DCP-1 expression (GAL) appears to totally prevent the formation of yeast cell colonies, i.e., it likely leads to cell death.

19b. DIAP1 coexpression with DCP-1 appears to suppress the killing activity of DCP-1. Colony formation occurs.

19c. HID or GRIM expression appears to counteract the inhibitory effect of DIAP1 on DCP-1 cell killing activity. Colony formation is prevented.

19d. DIAP1 appears to be an inhibitor of DCP-1 caspase activity. HID and GRIM either induce DCP-1 activity or interfere with the ability of DIAP1 to inhibit DCP-1 activity.

19e. DCP-1 and HID bind to GST-DIAP1, i.e., to DIAP1.

19f. As HID with the deletion of amino acids 1-36 did not bind to DIAP1, the HID domain for interacting with DIAP1 is likely to be in the first 36 amino acids of HID.

24

Cancer

PART A: *Chapter Summary*

Cancer is mis-regulation of cell multiplication caused mainly by somatic mutation, although inheritance also predisposes individuals to cancer development. Mutation can occur in a wide variety of proteins including growth factors and their receptors, signal transduction proteins, transcription factors, proteins that regulate apoptosis, and cell cycle regulatory proteins.

Both dominant gain-of-function mutations in proto-oncogenes and loss-of-function mutations in tumor suppressor genes are oncogenic. Alteration of tumor suppressor genes can result in loss of cell cycle checkpoints and programmed cell death. Proto-oncogenes are converted to oncogenes by point mutation or over-expression; overexpression can be caused by gene amplification, translocation, or integration of a slow acting retrovirus. The existence of transducing retroviruses led to the realization that oncogenes were variants of normal cellular genes, a critical concept in the molecular study of transformation. Changes in cell physiology due to oncoproteins include constitutive receptor activity in the absence of ligand, constitutive activity of signal transduction proteins, and loss of regulatory phosphorylation.

Normal chromosomes are required for regulated DNA synthesis and cell division. Individuals with inherited defects in DNA repair are sensitive to environmental factors and susceptible to a variety of cancers. Chromosomal abnormalities such as aneuploidy, translocations, and duplications are common in many malignancies. The role in oncogenesis of telomerase, the enzyme that repairs the ends of chromosomes, is being investigated. Chromosomes with damaged DNA trigger the gatekeeper activity of p53, a tumor suppressor protein, to prevent progression past G_1 of the cell cycle and replication of mutant genomes. Wild-type p53 is a transcriptional activator, causing expression of p21, an inhibitor of G_1 Cdk-cyclin complexes; it also promotes apoptosis. p53 is a tetramer with four identical subunits. Mutation of one allele inactivates all p53 oligomers, a dominant-negative phenotype, in contrast to other tumor suppressor genes for which mutation is recessive. Over 50% of human cancers exhibit p53 mutation. Since p53 is so central to oncogenesis, it is not surprising that proteins with which it interacts are also oncogenic. Mutant MDM2 and the transforming proteins of the DNA viruses SV40 and human papilloma virus are examples of such proteins.

The advances in elucidating the molecular mechanisms of cell transformation provide strategies for therapy. These results still must be correlated with the tumor's effect in the organism, for example, how tumors metastasize and how they recruit blood vessels to support their growth.

PART B: *Reviewing Concepts*

24.1 *Tumor Cells and the Onset of Cancer*
(lecture date _____)

1. Development of cancer is a multi-step process.

   a. How does information on the onset of cancer agree with this hypothesis?

   b. How did data from humans affected with colon cancer support this hypothesis?

   c. How does data from cell transformation support this hypothesis?

2. How do cancer cells differ from their normal counterparts with respect to growth, blood supply and life span?

24.2 *Proto-oncogenes and Tumor Suppressor Genes* (lecture date _____)

3. In referring to the war on cancer, J. Michael Bishop, a Nobel laureate for his work with oncogenes, quoted a line from the Pogo comic strip: "We have met the enemy and he is us." Why is this quotation appropriate?

24.3 *Oncogenic Mutations Affecting Cell Proliferation* (lecture date _____)

4. Although slow-acting retroviruses lack oncogenes, retroviral infection can activate proto-oncogenes, leading to oncogenesis.

   a. Describe the mechanism of proto-oncogene activation that can result from infection with a slow-acting retrovirus.

   b. In what other ways can proto-oncogenes be converted to oncogenes?

5. Some oncoproteins are altered forms of normal cellular proteins that regulate cell growth. For example, certain mutations in *her2* and *c-ras* convert these proto-oncogenes into oncogenes.

   a. What is the normal function of the protein encoded by each of these proto-oncogenes?

   b. In terms of function, how do the Neu and Ras oncoproteins differ from their normal counterparts?

6. The deletion of the gene for PTEN, a broad spectrum phosphatase, is seen in advanced stages of human cancers. How might its deletion contribute to cancer?

24.4 *Mutations Causing Loss of Cell-Cycle Control* (lecture date _____)

7. Once cells pass through the restriction point, they are committed to replicate DNA and enter the S phase of the cell cycle. The restriction point is regulated by cyclin D1, p16, and Rb. How does mutation in or overexpression of the genes encoding each of these proteins lead to loss of cell-cycle control and cancer?

24.5 *Mutations Affecting Genome Stability*
(lecture date _____)

8. The p53 protein was discovered through its association with SV40 T antigen and was assumed initially to be an oncoprotein.

   a. What is the current consensus as to the function of p53 and what evidence caused this change in view?

   b. How does the effect of mutation in the *p53* gene differ from the effect of mutation in the *RB* gene. What is the molecular basis for this difference?

9. Inhibition of cell death is one of the hallmarks of cancer. Cite two examples of how this occurs.

PART C: *Analyzing Experiments*

24.1 *Tumor Cells and the Onset of Cancer*

10. An enormous amount of epidemiological data on the incidence of cancer in the human population has been collected, analyzed, and organized for various purposes. Age incidence data from one standard

source, the International Agency for Research in Cancer, provides a graphic illustration that colon cancer is largely found in older people. What do these data indicate about the possible number of changes in cell genotype needed to induce development of colon cancer in humans?

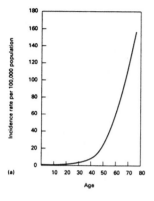
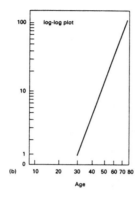

Figure 24-1

11. Experimental studies show that human leukemias and lymphomas overexpress both Myc and Bcl-2 proteins. How do these proteins cooperate to cause the cancer phenotype?

12. Plasminogen activators, which are secreted by many tumors, catalyze the conversion of plasminogen to plasmin, which in turn can activate other proteases that degrade components of the basal lamina. However, serum contains high levels of plasmin inhibitors, which theoretically can render plasmin inactive. Recently, workers have shown that fibrosarcoma cells have cell-surface receptors for both plasminogen activators and plasminogen/plasmin. Furthermore, plasmin bound to these cells is catalytically active and is known not to be inhibited by at least one of the serum plasmin inhibitors.

    a. Suggest a role for these proteases in tumor growth and progression.

    b. What do the observations concerning serum plasmin inhibitors and cell surface receptors on

fibrosarcoma cells suggest about the localization of the proteolytic processes?

24.2 Proto-Oncogenes and Tumor-Supressor Genes

13. Most transducing retroviruses contain the oncogene as a fusion protein with a normal retroviral gene. For example, the oncogene of avian myelocytosis virus-29 is named $p110^{gag-myc}$, where p stands for protein, 110 describes the size of the fusion protein in kDa, and gag-myc indicates a fusion with a structural protein. How do scientists maintain these viruses in the laboratory?

14. Acoustic neuroma is a human tumor derived from cells that surround the vestibulocochlear nerve. Although most cases of acoustic neuroma occur as unilateral, apparently non-inherited tumors, bilateral tumors are characteristic of an inherited form of this cancer. In order to understand the basis for the inherited disorder, DNA from leukocytes (normal cells) and tumor cells of patients with bilateral tumors were examined by Southern blotting with probes from known locations on human chromosomes. Hybridization of a labeled probe with two bands in the patient's DNA digests indicates heterozygosity in the patient's restriction fragments. The data from this study are shown in Table 24-1.

    a. Which chromosomal probes reveal differences in the leukocyte and tumor-DNA from patients? What is the most likely cause of these differences?

    b. What other cancer involves a similar type of alteration in the tumor cell DNA compared with normal DNA?

    c. What do these data suggest about the mechanism of oncogenesis in familial acoustic neuroma?

    d. In the tumor DNA from fourteen patients, no loss of heterozygosity was detected with the chromosome 22 probes. How might these data be explained?

Probe	Chromosomal location of probe	Number of patients with two leukocyte fragments that hybridize with probe	Number of patients with two leukocyte fragments that also have two tumor fragments that hybridize with probe
A	1	5	5
C	4	9	9
D	4	3	3
E	10	3	3
F	11	10	10
J	12	5	5
N	13	2	2
O	14	11	11
P	17	10	10
Q	18	9	9
R	19	2	2
U	22	6	3
V	22	12	8
W	22	4	3

Table 24-1

24.4 Tumor Cells and the Onset of Cancer, 24.4 Mutations Causing Loss of Cell-Cycle Control, 24.5 Mutations Affecting Genome Stability

15. Recently, the first conversion of human cells to cancer cells has been achieved. Transformation required the catalytic subunit of telomerase (hTERT), SV40 large T antigen, and an activated *ras* oncogene.

    a. One of the assays for transformation was the growth of colonies of cells in soft (low concentration) agar. Why is this a standard assay for transformation?

    b. Compare the number of components necessary for human cell transformation with those required to transform rodent cells.

    c. Other investigators have been unable to transform human cells with a combination of telomerase, *ras*, and the human papilloma virus E6/E7 oncoprotein. Why might this be the case?

    d. What are the differences between rodent and human cells that makes telomerase critical for human transformation?

Answers

1a. The incidence of a variety of cancers increases with age.

1b. Colon cancer progresses through a set of well-defined stages, each of which has been correlated with an additional mutation.

1c. Cell transformation requires (in rodents) at least two events. Recipient NIH/3T3 cells are mutated in p16, a cyclin kinase inhibitor that prevents progression through the cell cycle. Upon addition of the Ras oncoprotein, the cells become fully transformed. Similarly, an oncogene product located in the cytoplasm (e.g., a Ras protein) and one located in the nucleus (e.g., a Myc protein) are required to achieve complete transformation.

For further study, see text section "Tumor Cells and the Onset of Cancer," pp. 1057-1060 and Figures 24-4, 24-5, and 24-6.

2. Normal cells grow until they touch other cells, a phenomenon called contact inhibition. Mutant cells do not recognize the signals for contact inhibition; in tissue culture, they grow in mounds rather than flat sheets. Cancer cells actively recruit blood vessels for growth and metastasis, a process called angiogenesis. Normal cells have a defined life span in the organism and in tissue culture. An enzyme called telomerase can repair the ends of chromosomes; if it is not active, the telomeres shorten with each cell division, leading to chromosome fusion and death. Cancer cells produce telomerase, leading to immortalization.

For further study, see text sections "Tumor Cells and the Onset of Cancer," pp. 1056-1057 and "Mutations Affecting Genome Stability," p. 1081.

3. The Pogo quotation is apt because many human cancers result from alterations in cellular genes or metabolic processes. Once this was understood, researchers realized that retroviral oncogenes provided a model system for studying the structure and function of cancer-causing genes and their encoded proteins. With this approach, isolation of the relevant genes from the human genome, a difficult task, was not necessary.

For further study, see text section "Protooncogenes and Tumor Suppressor Genes," p. 1065.

4. Integration of a slow-acting retrovirus into the host-cell DNA can activate a cellular protooncogene. If the retroviral DNA is integrated upstream of the proto-oncogene, the promoter sequences in the right LTR will increase transcription of the proto-oncogene, which may be sufficient to cause transformation. This type of activation is called *promoter insertion*. If the integrated retroviral DNA is upstream in the reverse orientation or downstream of the proto-oncogene, the enhancer element in the LTR up-regulates transcription of the proto-oncogene by acting on the cellular promoter. This type of activation is called *enhancer insertion*.

For further study, see text section "Protooncogenes and Tumor Suppressor Genes," pp. 1065-1066 and Figure 24-10.

5a. The Her2 protein is a cell-surface receptor with protein tyrosine kinase activity, and Ras is an intracellular signal transducer with GTPase activity that is coupled to receptor tyrosine kinases.

5b. Both proteins normally respond to signals in an "on-off" manner, whereas their corresponding oncoproteins (Neu is the oncogenic form of Her2) are consistently "on."

For further study, see text section "Oncogenic Mutations Affecting Cell Proliferation," pp. 1070-1073 and Figure 24-15.

6. PTEN might affect the level of phosphorylation of tyrosine in proteins. The activity of certain oncoproteins such as Src is regulated by the presence or absence of phosphate on a particular tyrosine residue. Some growth-factor receptors are protein tyrosine kinases, whose activity might be affected by loss of a phosphate group. It is suggested that PTEN actually affects apoptosis and cell cycle progression.

For further study, see text section "Oncogenic Mutations Affecting Cell Proliferation," p. 1073.

7. At the restriction point of a normal cell cycle, p16, cyclin D1, and its associated Cdks are bound together in an inactive complex. At this time, Rb is hypophosphorylated and prevents activity of *E2F* transcription factors responsible for the eventual synthesis of enzymes required in S phase, such as DNA polymerase. Overexpression of cyclin D genes can occur in cancers through duplication or

translocation to a different chromosome, where they become subject to the regulation of an inappropriate promoter. Such overexpression can override the restriction point. Mutation in p16 inhibits its ability to inactivate the cyclin D1-cdk complex. Mutation in Rb or active/overactive cyclin D1-cdk complex phosphorylates Rb, permitting transcriptional activation.

For further study, see text section "Mutations Causing Loss of Cell-Cycle Control," pp. 1074-1075.

8a. p53 is believed to be a tumor-suppressor protein. This view is based on the finding that many tumors contain mutations, deletions, or rearrangements in the *p53* alleles. A loss-of-function mutation defines a tumor-suppressor gene.

8b. Mutation of even one *p53* allele is a dominant mutation, in contrast to mutation in *RB*. Both *RB* alleles must be mutated for induction of retinoblastoma. This difference is caused by the fact that the p53 protein acts as an oligomer, and the presence of even one defective subunit in the complex abrogates its function. In contrast, the Rb protein acts as a monomer.

For further study, see text section "Mutations Affecting Genome Stability," pp. 1076-1078 and Figure 24-21.

9. Inhibition of cell death can be caused by expression of Bcl-2, an inhibitor of apoptosis, or by expression of telomerase. Burkitt's lymphoma is due to Bcl-2 overexpression caused by its translocation to a chromosomal site where it is under the regulation of an antibody enhancer. Telomerase may prevent normal cell death due to its ability to repair the ends of chromosomes and thereby overcome senesence.

For further study, see text section "Mutations Affecting Genome Stability," pp. 1079-1081.

10. Clearly, the incidence of colon cancer does not increase in a linear manner with age, as would be expected if in a single-gene or an epigenetic change. The relationship between incidence and age is approximately linear when plotted as a log-log plot. The slope of the log-log plot is 5. This is a result consistent with multiple changes being needed to produce cancer in humans. In fact, the data suggest that six independent changes are required to produce colon cancer.

For further study, see text section "Tumor Cells and the Onset of Cancer," pp.1059-1060 and Figure 24-6.

11. The Myc protein is a transcription factor promoting expression of proteins involved in the cell cycle. A cell overexpressing Myc undergoes apoptosis, presumably because it senses an internal growth signal without receiving a complementary growth signal from the cell surface. This apoptosis is overcome by the activity of Bcl-2.

For further study, see text section "Tumor Cells and the Onset of Cancer," p. 1061.

12a. Together these proteases may allow tumor cells to penetrate the basal lamina, capillary walls, and interstitial connective tissue, so that the tumor can spread and establish metastases.

12b. Because of the receptors, plasminogen may be converted to plasmin on the surface of tumor cells. The resulting bound plasmin may be unaffected by high levels of serum plasmin inhibitors. Plasminogen/plasmin localized to tumor-cell surfaces, in addition to soluble enzymes, may contribute to tumor invasiveness.

For further study, see text section "Tumor Cells and the Onset of Cancer," p. 1056.

13. Most transducing retroviruses are defective, with the oncoprotein produced as a fusion protein with a portion of a normal retroviral protein (Gag, Pol, or Env). In order to propagate a transducing retrovirus, cells must be co-infected with a helper virus, e.g., the nontransforming variant of the transducing virus that would supply the missing or defective viral proteins.

For further study, see text section "Protooncogenes and Tumor Suppressor Genes," p. 1065.

14a. Probes U, V, and W reveal that the DNA of tumor cells from some patients has only one restriction fragment that hybridizes with markers on chromosome 22, whereas the leukocyte DNA from the same patients has two different restriction fragments hybridizing with these markers. This finding suggests that the tumor cells of these individuals have lost at least some of the DNA from one of their copies of chromosome 22.

14b. Retinoblastoma involves a small deletion on chromosome 13.

14c. These data suggest that a deletion on chromosome 22 is oncogenic. Acoustic neuroma, like retinoblastoma, appears to be brought about by a relatively rare somatic cell mutation, making the cell homozygous recessive. (The inherited defect is the presence of one recessive allele, indicated by the observed heterozygosity in leukocytes.) The fact that the loss of the second allele causes cancer suggests that the wild-type gene functions as a tumor suppressor gene.

14d. The lack of heterozygosity in some patients has several possible explanations: 1. The tumors in these patients may have been induced by some mechanism that does not involve the loss of genes on chromosome 22. 2. These tumor cells may have the same defective allele as those of other patients, but the change that occurred in the normal allele of the tumor-associated gene on chromosome 22 may not have altered the size of the restriction fragment, e.g. a point mutation. 3. A deletion may have occurred in the normal allele but it did not include the DNA in the restriction fragments that hybridized to the probes.

For further study, see text section "Protooncogenes and Tumor Suppressor Genes," p. 1067.

15a. The ability of cells to grow in soft agar mimics the anchorage-independent growth of transformed cells in culture and the ability of cells to metastasize.

15b. Transformation of rodent cells requires two components whereas transformation of human cells requires three components. In human cells, four targets are involved, since SV40 large T antigen interacts with both Rb and p53.

15c. Several reasons are suggested for the difference in results. The first is that SV40 large T antigen may affect cell constituents other than p53 and Rb in as yet unknown ways that contribute to immortalization. The other may simply be the order in which the cells are exposed to these genes. The successful experiments exposed cells first to SV40 large T antigen while the unsuccessful experiments first added hTERT.

15d. Rodent cells have longer telomeres and express higher levels of telomerase than human cells. Therefore, there would not be as great a need for upregulation of telomerase to immortalize rodent cells.

For further study, see text sections "Tumor Cells and the Onset of Cancer," pp. 1058-1059; and Figure 24-4, "Mutations Causing Loss of Cell-Cycle Control," p. 1075; and "Mutations Affecting Genome Stability," p. 1081.

APPENDIX

Complete Solutions
to *Testing Yourself on the Concepts*
and *MCAT/GRE-Style Questions*

Chapter 2. Chemical Foundations

Testing Yourself on the Concepts

1. The conversion of ADP to ATP has a standard free energy of 7.3 kcal. Under aerobic conditions, 36 moles of ATP is generated per mole of glucose (263 kcal/7.3 kcal per mole of ATP). Under anaerobic conditions, 2 moles of ATP is generated per mole of glucose (14.6 kcal/7.3 kcal per mole of ATP). The total energy yield from the oxidation of glucose is 686 kcal. The metabolism of glucose to lactic acid captures only 2.1% of this (14.6 kcal/686 kcal × 100).

2. At the acid pH of a lysosome, ammonia is converted to ammonium ion. Ammonium ion is unable to traverse the membrane because of its positive charge and is trapped within the lysosome. The accumulation of ammonium ion decreases the concentration of protons within lysosomes and therefore elevates lysosomal pH. At neutral pH, ammonia has little, if any, tendency to protonate to ammonium ion and, so, has no effect on cytosolic pH.

3. The introduction of a double bond, unsaturation, introduces a kink into a fatty acyl chain. Because of the kink, the unsaturated chain cannot pack as well into a rigid order with other fatty acyl chains. Hence as temperature decreases, *E. coli* maintains a relatively, disordered fluid state to its membrane by decreasing the ratio of saturated to unsaturated fatty acyl chains (increasing the number of unsaturated fatty acyl chains relative to saturated ones).

4. At least three properties contribute to this structural diversity. First, monosaccharides can be joined to one another at any of several hydroxyl groups. Second, the C-1 linkage can have either an α or a β configuration. Third, extensive branching of carbohydrate chains is possible.

MCAT/GRE-Style Questions

1. a; 2. c; 3. d; 4. b; 5. b; 6. b; 7. a; 8. d; 9. a; 10. c; 11. d; 12. b; 13. c; 14. a

Chapter 3. Protein: Structure and Function

Testing Yourself on the Concepts

1. The structure of proteins is commonly described in terms of four hierarchical levels of organization. The *primary* structure is the linear arrangement, or sequence, of amino acids that constitute the polypeptide chain. The *secondary* structure refers to specialized structures, for example α helices or β sheets, that are formed by regions of the polypeptide chain. These structures are stabilized or held together by hydrogen bonds. The *tertiary* structure refers to the overall conformation of the polypeptide, or three-dimensional arrangement of all the amino acids.

The tertiary structure is stabilized by hydrophobic interactions between nonpolar side chains and, in some cases, disulfide bonds between cysteine residues. The *quaternary* structure describes the number and organization of polypeptide chains or subunits in a multimeric protein. The subunits are held together by noncovalent bonds.

2.   An enzyme is a protein that increases the rate of a reaction by lowering the activation energy and stabilizing transition-state intermediates. An enzyme contains an active site, which comprises two functional parts: a substrate-binding region and a catalytic region. The amino acids that make up the active site do not need to be adjacent in the polypeptide chain, but in fact can be from different regions of the chain that are brought together by protein folding. A cellular enzyme must be able to catalyze its reaction in an environment normally encountered in a cell, i.e., in aqueous solution, at pH 6.5–7.5, at 37°C. A cellular enzyme may also be an allosteric enzyme. An allosteric enzyme would have not only a binding site for the substrate (the active site) but also one or more binding sites for effector molecules, which can alter the activity of the enzyme.

3.   Membrane proteins can be classified into two broad categories, *integral* and *peripheral*, based on the nature of the protein-membrane interactions. Integral membrane proteins, also called *intrinsic proteins*, have one or more segments that are embedded in the phospholipid bilayer of the membrane. These membrane-spanning segments exist as α helices or multiple β strands and contain hydrophobic amino acids that interact with the hydrophobic fatty acyl groups of the membrane phospholipids. Some integral membrane proteins do not actually span the phospholipid bilayer but are covalently attached to fatty acids that are embedded in the membrane. Peripheral membrane proteins do not span the phospholipid bilayer but instead are bound to the membrane indirectly by interactions with integral membrane proteins or directly by interactions with the polar head groups of phospholipids.

4.   Proteins can be separated on the basis of their mass or charge. Methods used for separating proteins based on their mass include rate zonal centrifugation, SDS-gel electrophoresis, and gel filtration chromatography. In rate zonal centrifugation, proteins are centrifuged through a solution, usually sucrose, of increasing density. Large proteins migrate faster through this density gradient and can be separated from smaller proteins. In SDS-gel electrophoresis, proteins are first treated with the ionic detergent sodium dodecylsulfate (SDS). SDS denatures proteins and forces proteins into extended conformations with similar charge:mass ratios. SDS treatment masks the charges of the amino acid side groups and allows proteins to be separated on the basis of chain length, which is a reflection of mass. Small proteins migrate more rapidly toward the positive pole in a polyacrylamide gel. In gel filtration chromatography, proteins flow around porous beads made from polyacrylamide, dextran, or agarose packed into a column. Smaller proteins can penetrate the beads more easily, and thus travel more slowly through a gel filtration column compared with larger proteins. As a result, large proteins exit the column first.

Methods used for separating proteins on the basis of their charge include isoelectric focusing and ion-exchange chromatography. In isoelectric focusing, proteins are separated by electrophoresis using polyacrylamide saturated with ampholytes, which are a mixture of positively and negatively charged molecules. During electrophoresis the ampholytes separate and form a pH gradient. The proteins migrate through this gradient until they reach their isoelectric point, or the pH at which the net charge on the protein is zero. In ion-exchange chromatography, charged proteins can bind to beads whose surfaces carry the opposite charge. For example, negatively charged proteins, but not neutral or positively charged proteins, will bind to a positively charged bead. Bound proteins can then be eluted off the bead by a gradient of increasing salt concentration. Weakly charged proteins are eluted first and highly charged proteins are eluted last.

MCAT/GRE-Style Questions

1. b; 2. d; 3. a; 4. d; 5. b; 6. d; 7. c; 8. c; 9. b; 10. c; 11. a; 12. c

Chapter 4. Nucleic Acids, the Genetic Code, and the Synthesis of Macromolecules

Testing Yourself on the Concepts

1. If one does not have to consider the enzymes required for synthesis or replication of DNA itself, the minimum components are RNA polymerase required for transcription of DNA into RNA; rRNAs; ribosomes; tRNAs; amino acids; aminoacyl-tRNA synthetases; and initiation, elongation, and termination factors. If one is considering information flow in a eukaryote, one would have to add the machinery to cap, polyadenylate, and splice mRNAs, and transport them into the cytosol.

2. Primers must base-pair with a specific strand of DNA, so they must contain a complementary sequence in order to anneal. The T_m of each primer is used to determine the annealing temperature of the reaction. The higher the temperature of the reaction, the more specific the amplification is; however, the reaction temperature must not exceed the T_m of the primers or annealing will not occur. Lastly, as in DNA replication, each oligonucleotide primer must be able to incorporate new deoxyribonucleotides in a 5′-to-3′ manner; therefore, primers must have a free 3′-hydroxyl end.

3. The mutation may have occurred outside the coding region. For example, mutations in introns or 5′ and 3′ UTRs might not affect transcription of the gene, and would not affect the translation of the protein. Alternatively, not all mutations in the coding region result in loss of gene function. If the single base change occurred in the wobble position of a codon, an amino acid substitution would not necessarily occur. If the single base change did result in an amino acid substitution, this substitution might not alter the final conformation of the protein, thus allowing it to function normally. Lastly, a single base change resulting in a new stop codon might not affect the function of a protein if this occurred near the carboxyl terminus and the deleted region was not needed for activity.

MCAT/GRE-Style Questions

1. d; 2. b; 3. d; 4. a; 5. c; 6. a; 7. a; 8. c; 9. a; 10. d; 11. c

Chapter 5. Biomembranes and the Subcellular Organization of Eukaryotic Cells

Testing Yourself on the Concepts

1. Electron microscopy has better resolution than light microscopy, but many light-microscopy techniques allow observation and manipulation of living cells.

2. Specific types of cells in suspension may be isolated by a fluorescence-activated cell sorter (FACS) machine in which cells previously "tagged" with a fluorescent-labeled antibody are separated from cells not recognized by the antibody. The scientist selects an antibody specific for the cell type desired. Specific organelles are generally separated by centrifugation of lysed cells. A series of centrifugations of successive supernatant fractions at increasingly higher speeds and corresponding higher forces serve to separate cellular organelles from one another on the basis of size and mass (larger, heavier cell components pellet at lower speeds). This is often combined with density-gradient separations to purify specific organelles on the basis of their buoyant density.

3.   The amphipathic nature of phospholipid molecules (a hydrophilic head and hydrophobic tail) allows these molecules to self-assemble spontaneously into closed bilayer structures when in an aqueous environment. The phospholipid bilayer provides a barrier with selective permeability that restricts the movement of hydrophilic molecules and macromolecules in and out of the compartment. The different types of proteins present on the two faces of the bilayer contribute to the distinctive functions of each compartment's interior and exterior, and control the movement of selected hydrophilic molecules and macromolecules across the bilayer.

4.   The multiple membranes of mitochondria and chloroplasts act to create additional compartments with specialized functions within these organelles. The polarized stack of compartments that form the Golgi apparatus is associated with the assembly line organization of enzymes that modify many ER-derived products.

MCAT/GRE-Style Questions

1. c; 2. b; 3. b; 4. a; 5. c; 6. c; 7. a; 8. b; 9. a; 10. d; 11. c; 12. b; 13. d

Chapter 6. Manipulating Cells and Viruses in Culture

Testing Yourself on the Concepts

1.   They are a clone of cells, so that they are identical; they are immortal, so they grow forever in culture; they produce the monoclonal antibody characteristic of their B-lymphocyte progenitor.

2.   To isolate a double auxotroph such as this, one could mutagenize cells and then use the technique of replica plating to search first for mutants that grow on minimal medium containing leucine but not on minimal medium deficient in leucine. The procedure would be analogous to that shown in Figure 6-2, which depicts the isolation of bacteria deficient only in arginine synthesis. A clone of such cells is isolated, grown to large numbers, subjected to mutagenesis, and then replica-plated to identify a clone that can grow on minimal medium plus leucine and adenine, but not on minimal medium lacking either leucine or adenine.

3.   The classification scheme is based on the mechanism of viral mRNA synthesis and the "strandedness" (+, −, or double-stranded) of the viral genome, rather than on the type of viral host, shape of the virion, or nature of the viral proteins (see Figure 6-20).

MCAT/GRE-Style Questions

1. c; 2. c; 3. b; 4. b; 5. d; 6. a; 7. c; 8. a; 9. b; 10. b; 11. b; 12. d; 13. a; 14. a; 15. d

Chapter 7. Recombinant DNA and Genomics

Testing Yourself on the Concepts

1.   Genomics allows scientists to obtain information on the presence or absence of metabolic pathways, the functions of proteins encoded by the genome, and the proportion of the genome devoted to specialized functions. Studies of the archaeans show that, for certain essential functions such as replication, transcription, and translation, they are more related to eukaryotes than to prokaryotes. *S. cerevisiae* has many proteins devoted to secretion and membrane targeting,

reflecting the presence of the nucleus and cytoplasmic organelles in these cell. The *C. elegans* genome displays yet another level of complexity, with proteins devoted to the cell-cell interactions required in a multicellular organism.

2.   The two classes of enzymes that allow cloning of DNA are restriction endonucleases and DNA ligase. Each restriction endonuclease recognizes a particular DNA sequence in any DNA in which the sequence occurs. This property, along with the ability of DNA ligase to join pieces of DNA, means that any two pieces of DNA with similar ends can be joined together. For detailed analysis of a particular DNA sequence, it is joined to DNA that can be obtained in large amounts. The enzyme that facilitates DNA sequencing is polynucleotide kinase; it adds a detectable molecule to the 5' end of the primer that initiates DNA synthesis in the Sanger method. Without this label, the size of the fragments produced cannot be determined. PCR was automated through the use of a thermostable DNA polymerase. Such a polymerase is not inactivated at the temperatures required to melt DNA, so that multiple rounds of primer hybridization and DNA synthesis can occur. Reverse transcriptase, producing double-stranded DNA from a single-stranded RNA template, is used to prepare cDNAs, which are then cloned.

3.   Since the protein is of value, it is likely that it has been isolated and characterized. If an antibody to the protein were available, it could be used to screen a cDNA expression library. The cDNA would then be expressed in a system capable of producing large amounts of protein, such as the two-stage system in which the T7 RNA polymerase in induced to express proteins from the T7 promoter. Eukaryotic proteins that require certain post-translational modifications such as glycosylation cannot be produced in active form in bacteria; other systems derived from, for example, insect cells, can produce these proteins.

4.   The presence of a particular DNA or RNA in a mixture can be ascertained by electrophoresing the mixture and carrying out a Southern blot (for DNA) or a Northern blot (for RNA). For a Southern blot, the DNA is first digested with one or more restriction enzymes before electrophoresis. A particular DNA sequence can be cloned, and a bacterial colony or λ vector plaque containing only the cloning vector with an insert of the DNA of interest can be isolated. To isolate a unique mRNA, total RNA is prepared by oligo dT chromatography. The poly(A) tail on mRNA is used to hybridize an oligo dT primer to produce cDNA, and the cDNAs are cloned to produce a library; again, the cloning procedure results in each vector containing one cDNA.

MCAT/GRE-Style Questions

1. a; 2. b; 3. d; 4. a; 5. a; 6. b; 7. c; 8. b; 9. d; 10. d

Chapter 8. Genetic Analysis in Cell Biology

Testing Yourself on the Concepts

1.   Both mitosis and meiosis begin with a doubling of the chromosome complement to produce a cell with a 4*n* chromosome complement. Completion of mitosis requires one cell division, so that each daughter cell is 2*n*, while completion of meiosis requires two divisions, and each daughter cell is 1*n*. In mitosis, each pair of sister chromatids attaches to the spindle independently; the chromatids of each pair separate and move to opposite daughter cells. In meiosis, during the first meiotic division, homologous pairs of chromosomes align so that one pair of sister chromatids moves to one daughter cell, while the other pair goes to the other daughter cell. In the second meiotic division, the pairs of chromatids are separated, as in mitosis. Also, mitosis occurs in somatic cells, while meiosis occurs in specialized cells destined to become gametes.

2. Females from this cross will all be phenotypically wild type. Half the females will be heterozygous for the recessive mutation, and half will be homozygous for the wild-type allele. Half the males will be mutant, and half wild type. Assuming that all F1 progeny mate normally, seven-eighths of the female F2 progeny will be phenotypically wild type, and one-eighth will be mutant. Of seven wild-type females, three are homozygous for the wild-type allele, and four are heterozygous. Of the male F2 progeny, three-quarters are wild type, and one-quarter are mutant.

3. Using knockout mice as models for human diseases requires the isolation of the human gene whose mutation results in the condition and the isolation of the same gene in mice. The mouse gene is then mutated specifically in vitro and introduced into mice by the techniques illustrated in Figures 8-33, 8-34, 8-35, and 8-36. It is predicted that the phenotype of mice homozygous for the mutation will be the same as that of humans with the disease. The physiological basis for the disease can then be studied in the mutant mice.

4. The principle behind using recombination frequency to order genes on chromosomes is that the farther away two genes are from each other, the more likely recombination will occur (see Figure 8-18).

MCAT/GRE-Style Questions

1. b; 2. c; 3. b; 4. d; 5. b; 6. a; 7. d; 8. a; 9. d; 10. d; 11. b; 12. b; 13. b

Chapter 9. Molecular Structure of Genes and Chromosomes

Testing Yourself on the Concepts

1. Genes can be classified as simple or complex transcription units. A simple transcription unit encodes a single mRNA, which encodes a single protein. A mutation in an exon of a simple transcription unit affects only one protein. A complex transcription unit encodes an mRNA that can be processed in more than one way by the use of alternative poly(A) sites or splice sites. A transcription unit that contains two or more poly(A) sites can produce mRNAs that have the same 5′ exons but different 3′ exons. Another variation of alternative RNA splicing, also known as exon skipping, can generate mRNAs with the same 5′ and 3′ exons but different combinations of internal exons. These mRNAs with various exon combinations likely encode proteins with different amino acid sequences. Mutations in an alternatively spliced exon will affect only those proteins that contain the mutated exon. One benefit to complex transcription units is that they provide a mechanism for generating a number of different proteins from a single gene.

2. Mobile elements can be classified into two categories based on their mode of transposition. Mobile elements that transpose directly as a DNA intermediate are referred to as transposons, whereas mobile elements that transpose through an RNA intermediate are called retrotransposons. An insertion sequence (IS) element contains an inverted repeat at each end of the insertion sequence. Between the inverted repeats is a sequence that encodes a transposase enzyme, which mediates that transposition event. During nonreplicative transposition, the transposase excises the IS element from one location and inserts it into a new position. Some IS elements transpose by a replicative process. During this process, a copy of the IS element is synthesized and inserted into a new position while retaining the original IS element. Bacterial transposons are typically composed of an antibiotic resistance gene flanked by two copies of the same type of IS element. For both IS elements and transposons, transposition results in the duplication of the target site on either side of the inserted mobile element.

Retrotransposons transpose through an RNA intermediate and are subdivided into either viral or nonviral retrotransposons. Viral retrotransposons contain short 5′ and 3′ direct repeats and 250- to 600-bp long-terminal repeats (LTRs) flanking the central protein-coding region. The leftward LTR acts as a promoter for transcription of the protein-encoding genes. These proteins, reverse transcriptase and integrase, are necessary for converting the RNA intermediate into DNA and then insertion of the DNA into the target site. In contrast, the mechanism mediating transposition of nonviral retrotransposons, which lack LTRs, is less well understood. Promoter sequences in the left side of the element direct transcription of the retrotransposon, including the encoded reverse transcriptase gene. The reverse transcriptase synthesizes a DNA copy, and through some mechanism the retrotransposon is inserted into its target site. Examples of nonviral retrotransposons include the 6- to 7-kb long interspersed elements (LINES) and 300-bp short interspersed elements (SINES).

3. Antibodies consist of two identical heavy chains (55 kDa) and two identical light chains (23 kDa). Both heavy and light chains are composed of variable and constant domains. The DNA region containing the kappa light chains genes consists of about 100 variable (V) segments and five joining (J) segments. Random joining of one variable segment with one joining segment by a site-specific recombinase can produce 500 different combinations. Even greater diversity is generated due to the imprecise joining of V and J segments. A small, variable number of nucleotides are lost or added during V-J joining, resulting in different amino acid sequences at the V-J junction. A similar mechanism exists for generating diversity of heavy chain genes. The variable region of heavy chain genes consists of not only 100 variable and 6 joining segments but also 30 diversity (D) segments. Random joining of the V, D, and J segments can produce 18,000 different combinations. Again, random loss or addition of a variable number of nucleotides at the V-D-J junctions generates even greater diversity. Because an immunoglobulin molecule consists of both a heavy and light chain, then 9 million (500 × 18,000) different antibody molecules could be made. Somatic mutation, which was not discussed in this chapter, is yet another mechanism for generating further antibody diversity.

4. Metaphase chromosomes contain multiple levels of organization and folding of the DNA duplex. A segment of the DNA duplex is first wrapped around a histone octamer core to form a nucleosome. This histone:DNA complex is known as chromatin and resembles a beads-on-a-string structure. This beaded structure undergoes further folding into a spiral or solenoid arrangement forming a 30-nm chromatin fiber. These 30-nm fibers attach to a flexible protein scaffold at intervals of millions of base pairs resulting in long loops of chromatin extending from the scaffold. Coiling of this scaffold:DNA complex into a helix and further packaging of this helical structure produces a highly condensed structure characteristic of metaphase chromosomes.

Three structural elements that are necessary for replication and stable inheritance of chromosomes are origins or replication, the centromere, and two telomeres. Origins of replication or autonomously replicating sequences (ARSs) are sequences in the DNA, which are sites for the initiation of DNA synthesis. Centromeres are short, relatively well-conserved sequences that are required for proper segregation of chromosomes. Centromeres attach to spindle microtubules during mitosis and meiosis (see Chapter 19). Telomeres are specialized structures located at the ends of linear chromosomes. The DNA sequences of telomeres consist of repetitive short oligomers repeated a few hundred to a few thousand times. These telomeric repeats are added by an enzyme known as telomerase.

MCAT/GRE-Style Questions

1. b; 2. d; 3. a; 4. d; 5. b; 6. a; 7. b; 8. c; 9. a; 10. c; 11. d; 12. b

Chapter 10. Regulation of Transcription Initiation

Testing Yourself on the Concepts

1. When *E. coli* grows in a medium lacking lactose, the enzymes required for lactose metabolism encoded in the *lac* operon are not expressed to a significant level. The *lac* operon is repressed by the protein product encoded by the *lacI* gene, the *lac* repressor. The *lac* repressor binds to the operator sequence of the *lac* operon, blocking transcription of the operon by RNA polymerase. In medium containing lactose, the enzymes of the *lac* operon are induced. Lactose enters the cell and binds to the *lac* repressor, causing a conformational change in the shape of the repressor molecule that prevents the repressor from binding to the operator. RNA polymerase is then free to bind to the promoter sequence and initiate transcription of the *lac* operon (see Figure 10-2).

2. Two methods that are commonly used to examine DNA-protein interactions are the DNase I footprinting assay and the electrophoretic mobility shift assay (EMSA). In the DNase I footprinting assay, the DNA fragment to be analyzed for protein-binding sites is first radioactively labeled at one end. The protein is allowed to bind to the radiolabeled DNA, and then the DNA is partially digested with DNase I. The DNase I digestion conditions are set so that on average each DNA molecule is cleaved once. The DNA fragments are then denatured and separated on a gel. The DNA region bound by the protein is inaccessible to DNase I digestion and thus appears as a gap or footprint on the autoradiogram (see Figure 10-6). In the EMSA, the DNA fragment to be analyzed for protein-binding sites is first radioactively labeled. The protein is allowed to bind, and then the DNA is separated on a nondenaturing gel. Binding of a protein to the DNA fragment reduces the fragment's mobility, causing a shift in its location, which is detected by autoradiography (see Figure 10-7).

    The location of the DNA control elements in a regulatory region can be determined by deleting various regions of the regulatory region (5′-deletion analysis) or inserting scrambled DNA sequences into the regulatory region (linker scanning mutation analysis). In 5′-deletion analysis, DNA fragments containing various lengths of the sequence upstream of the transcription start site are cloned in front of a reporter gene. These regulatory element-reporter gene constructs are transfected into cells, and reporter gene activity is measured in cell extracts. The approximate locations of important regulatory elements are determined by the relative level of reporter gene activity of the DNA constructs (see Figure 10-24). In the linker scanning mutation assay, a systematic set of constructs with overlapping mutations are cloned in front of a reporter gene. These constructs are transfected into cells, and activity of the reporter gene is assayed in cell extracts (see Figure 10-31). In both assays, the correlation between reporter gene activity and position of the 5′ deletion or location of the scrambled sequence provides an approximate location of the regulatory element.

3. In prokaryotes a single RNA polymerase transcribes all RNA; whereas in eukaryotes, RNA polymerase II transcribes mRNA. The major form of bacterial RNA polymerase is composed of five subunits: β, β′, 2 copies of α, and one copy of σ^{70}. Transcription initiation begins when the σ subunit interacts with promoter sequences near -10 and -35 base pairs from the transcription start site. The polymerase complex synthesizes approximately 10 nucleotides, and then the σ subunit is released.

    In eukaryotes, transcription usually begins at a TATA box located 25–35 base pairs upstream of the transcription start site or at an initiator element. Transcription initiation on genes that contain a TATA box begins with binding of TFIID, which contains the TATA box-binding protein (TBP). Other transcription factors (TFIIB, TFIIF, TFIIE, and TFIIH) and RNA polymerase then subsequently bind to form a large DNA-protein complex. As the polymerase begins transcribing away from the start site, the carboxyl-terminal domain of RNA polymerase II is phosphorylated.

4.   Transcriptional activator and repressor proteins consist of two functional domains: a DNA-binding domain and an activation or repression domain. Remarkably, these domains can act independently; for example, a DNA-binding domain can be transferred to a protein that does not normally bind DNA to produce a protein that now binds DNA. The DNA-binding domains exhibit a variety of structures. Some of the most common structures are the homeodomain, basic zipper (leucine zipper), helix-loop-helix, and zinc fingers. Activation domains exhibit considerable structural diversity. Although a number of diverse amino acid sequences can act as activation domains, many activation domains are rich in acidic amino acids (aspartic acid and glutamic acid) and are called *acidic activation domains*. Likewise, a variety of amino acid sequences can function as repression domains. Some of these repression domains contain a high proportion of hydrophobic amino acids, whereas others contain a high proportion of basic amino acids.

MCAT/GRE-Style Questions

1. c; 2. d; 3. a; 4. d; 5. b; 6. d; 7. a; 8. b; 9. a; 10. c; 11. b; 12. c

Chapter 11. RNA Processing, Nuclear Transport, and Post-Transcriptional Control

Testing Yourself on the Concepts

1.   Differential splicing can create different mRNAs containing alternative coding regions. Regulation of mRNA stability can control the amount of encoded protein produced. Regulation of translation initiation can also control the amount of encoded protein produced. In some organisms, regulation of trans-splicing controls the synthesis of certain mRNAs. In the case of U1A mRNA, polyadenylation is regulated to control the amount of messenger RNA produced. In bacteria, synthesis of antisense RNA regulates the translation of some mRNAs.

2.   If a genomic copy of the human gene is used as a template for RNA transcription, splicing of introns will not occur. To provide an mRNA that can be translated in vitro, a cDNA clone (that contains no intron sequences) must be used as the template for mRNA transcription. Since a purified bacteriophage RNA polymerase is used to transcribe the cDNA clone, the transcribed RNA will not contain a 5′ cap unless transcription is initiated with a cap analogue. This is generally done by adding a high concentration of GpppG to the in vitro transcription reaction. Under these conditions, this cap analog is used to initiate transcription from a bacteriophage T7 RNA polymerase promoter. Since synthesis of a full-length cDNA clone is generally primed with oligo-dT (see Figure 7-15), the cDNA clone usually contains a poly-dA/poly-dT sequence at its 3′ end. This sequence is transcribed into a poly(A) tail during transcription of the cDNA clone by T7 RNA polymerase.

3.   All three types of eukaryotic RNA are synthesized as precursors that undergo modifications to yield mature RNAs. In all three types of RNA, portions are removed. Introns are spliced out of mRNA precursors. Introns are also spliced out of rRNA precursors in some organisms. Short introns are also spliced out of the anticodon loop in some tRNA precursors, but the mechanism of splicing differs from the general mechanism used in pre-mRNA and pre-rRNA splicing. Pre-tRNA splicing is catalyzed by proteins, whereas pre-mRNA and pre-rRNA splicing is catalyzed by ribozymes. The 5′ end of a nascent pre-mRNA is capped by enzymes that associate with the phosphorylated CTD of RNA polymerase II. The 3′ end of the transcribed portion of an mRNA is generated by cleavage at a poly(A) site. A residues are then polymerized onto the 3′ end generated by cleavage to produce the mRNA poly(A) tail. In rRNA, cleavage of a long pre-rRNA results in the production of 28S, 18S, and 5.8S rRNAs. Specific bases and riboses are methylated as directed by snoRNAs complementary to the regions of methylation. The 5′ ends of the pre-tRNAs

are cleaved by the ribosyme in RNase P. The mature 3′ end of a tRNA is generated by cleavage of the pre-tRNA followed by addition of CCA. Specific tRNA bases are subsequently modified by specific enzymes for each modification.

4. The nucleus normally does not allow unprocessed mRNAs to be exported in that pre-mRNAs associated with snRNPs in spliceosomes are prevented from being transported to the cytosol. However, 9-kb and 4-kb HIV RNAs containing splice sites are exported to the cytosol in HIV-infected cells. Mutants in the HIV Rev protein prevent this export; therefore, Rev must be involved in the export of these unspliced RNAs. Also, if the Rev-binding site, RRE, is mutated, then the 9-kb and 4-kb HIV RNAs are not exported. Thus, Rev binding to the RRE sequence is required for Rev-mediated transport.

MCAT/GRE-Style Questions

1. d; 2. a; 3. f; 4. f; 5. a; 6. e; 7. d; 8. e; 9. d, 10. d; 11. c

Chapter 12. DNA Replication, Repair, and Recombination

Testing Yourself on the Concepts

1. The mechanism for DNA replication in prokaryotes is similar to that in eukaryotes. In *E. coli*, the first step is the binding of DnaA protein to the origin, oriC. DnaB, which is a helicase, binds next and melts the duplex DNA. The resulting single strands are prevented from reannealing by the binding of single-strand binding protein. *E. coli* primase binds and catalyzes the synthesis of RNA primers for the synthesis of Okazaki fragments by DNA polymerase III. The β subunit binds as a dimer to polymerase III and functions as a clamp to keep the polymerase associated with the DNA template. As the newly synthesized Okazaki fragment approaches the 5′ end of the adjacent Okazaki fragment, DNA polymerase III dissociates and DNA polymerase I binds and removes the RNA primer and fills in the gap. Finally, DNA ligase joins adjacent lagging strand fragments.

    In eukaryotes, a similar process occurs. SV40 can serve as a model eukaryotic system. A viral protein called T antigen binds to the origin and using its helicase activity melts the duplex DNA similar to the *E. coli* DnaB helicase. Replication protein A then binds to the separated DNA strands like *E. coli* single-strand binding protein in. Polymerase α, tightly associated with primase, then binds. The activity of Pol α is stimulated by binding of replication factor C (RFC). Host-cell proliferating cell nuclear antigen (PCNA) binds and displaces the primase–Pol α complex. PCNA acts in an analogous fashion to the β-subunit clamp associated with DNA polymerase III. Both increase the processivity of DNA polymerase. Polymerase δ binds to the PCNA/RFC complex and completes DNA synthesis.

2. A DNA molecule that contains base-pairing errors or base damage can be repaired by a number of processes. During synthesis the proofreading function of DNA polymerase III removes incorrect base pairs. A 3' → 5' exonuclease removes the mismatched base and then DNA synthesis proceeds. Damage to DNA can be repaired by a number of mechanisms such as mismatch repair, excision repair, and double-strand break repair. Mismatch repair of single base mispairs is dependent upon the hemimethylation state of newly replicated DNA. In *E. coli*, DNA is methylated at the adenine in GATC sequences by the enzyme Dam methylase. The parental strand is methylated, whereas the newly synthesized daughter strand is temporarily unmethylated. Thus the methylation state of the DNA strands can serve as a marker to discriminate between the parental and daughter strands. The MutHLS system recognizes the mismatched base pair and can discriminate between the parental and daughter strand. An endonuclease activity in the MutH protein specifically binds to the DNA and cleaves the unmethylated daughter strand. The misincorporated base is excised and replaced

A-10

with the correct base. In excision repair, the best understood system is the *E. coli* UvrABC system for removal of thymine-thymine dimers. A UvrA-UvrB complex binds to and scans the DNA helix for distortions in the helix. At a damaged base, UvrA dissociates and UvrC binds. UvrC has an endonuclease activity that cleaves the DNA on either side of the damaged site. The damaged fragment is removed by a helicase and degraded and the gap is filled in by DNA polymerase I and ligase. Repair of double-strand breaks involves end joining of nonhomologous DNAs. A complex of Ku protein and a protein kinase bind to the ends of the DNA molecules. The helicase activity of Ku unwinds both ends until a short region of homology between DNA molecules is exposed. The unpaired single-stranded ends are removed and the two DNA molecules are ligated. This process can introduce mutations because nucleotides are lost at the ends of the ligated molecules.

3.   Genetic recombination involves the exchange of DNA information between DNA molecules. The Holliday model describes the molecular events that lead to genetic recombination. In the first step, two homologous DNA molecules align. A nick is made in one strand of each of the DNAs. The two nicked strands then invade the other DNA and form duplexes in a process called strand exchange. At the site of the nicks the 3' and 5' ends are joined to produce a Holliday structure (see Figure 12-29); the branch point then migrates, creating a heteroduplex region. The Holliday structure can then assume two different configurations by isomerization of the structure. Depending upon which DNA strands are cut and rejoined, the resulting separated duplexes are either recombinant or nonrecombinant. A nonrecombinant molecule contains alleles to the left and right of the crossover site, all derived from the same initial DNA molecule, whereas a recombinant molecule contains alleles to the left derived from one DNA molecule and alleles to the right derived from the other DNA molecule.

MCAT/GRE-Style Questions

1. a; 2. c; 3. b; 4. a; 5. d; 6. d; 7. a; 8. c; 9. d; 10. c; 11. b; 12. c; 13. a

Chapter 13. Regulation of the Eukaryotic Cell Cycle

Testing Yourself on the Concepts

1.   The unidirectional and irreversible passage through the cell cycle is brought about by the degradation of critical protein molecules at specific points in the cycle. Examples are the proteolysis of the anaphase inhibitor at the beginning of anaphase, proteolysis of cyclin B in late anaphase, and proteolysis of the S-phase Cdk inhibitor at the start of S. The proteins are degraded by a multiprotein complex called a proteasome. This breakdown is signaled by the addition of multiple molecules of ubiquitin to one or more lysine residues in the target protein. The anaphase inhibitor and cyclin B are both polyubiquitinated by the APC complex. The S-phase Cdk inhibitor is polybiquitinated by the Cdc34 pathway.

2.   The wee phenotype in the yeast *S. pombe* displays smaller than usual cells. Premature entry into mitosis, before the cell has grown to the size that normally signals cell division, is the cause of this phenotype. Wee cells result from excess activity of Cdc2, the cyclin-dependent kinase of *S. pombe* MPF. Increased Cdc2 activity can result from mutation in *cdc2* (making it insensitive to Wee) or the *wee1* gene. Wee1 encodes a kinase that phosphorylates Tyr-15 on Cdc2. The presence of a phosphate group on this amino acid inhibits Cdc2 activity; therefore, mutation in *wee1* abolishes the inhibition, leading to overactivity. Alternatively, an excess of Cdc25, a phosphatase that removes the phosphate from Tyr-15, also leads to the wee phenotype.

3.   High-fidelity replication of chromosomal DNA is required for cell division. Synthesis of DNA occurs during S phase, after the cell passes through START or the restriction point. Prereplication

complexes that were assembled in G_1 are activated through phosphorylation by S-phase cyclin-Cdks. Phosphorylation initiates DNA synthesis and also prevents assembly of additional prereplication complexes, so that the genome is replicated only once each cell cycle. If the chromosomes are broken, a G_2 checkpoint prevents entry into mitosis.

During M phase, cyclin-Cdk complexes regulate the events of mitosis, chromosome condensation, spindle formation, and attachment of the chromosomes to spindle microtubules. If the cell senses that these events have occurred correctly, mitotic Cdk complexes activate the anaphase-promoting complex (APC) that targets the anaphase inhibitor for destruction and eventually targets mitotic Cdk complexes for destruction.

In the following G_1 phase, prereplication complexes and the APC are reassembled, although in inactive forms. There is an additional checkpoint for DNA integrity in G_1 to prevent damaged DNA from being replicated.

4.   Some cell-cycle proteins carry out their roles in regulating the cell cycle when phosphorylated and others are active when dephosphorylated. Examples of the former include the anaphase-promoting complex (APC); the *S. pombe* Cdc2, which is active when only threonine 161 is phosphorylated (not when doubly phosphorylated on threonine 161 and tyrosine 15); and the lamins, which depolymerize when phosphorylated. Phosphorylation of Rb breaks the Rb-E2F transcription factor heterodimer, releasing E2F to promote transcription of genes required for entry into S phase. In a variation on this mechanism, Sic1, an S-phase inhibitor, is marked for degradation by the cdc34 pathway only when it is phosphorylated, thereby permitting initiation of DNA synthesis.

Dephosphorylation of lamins is required for reassembly of the nuclear envelope after mitosis. Dephosphorylation of inhibitory sites on myosin light chain allows cytokinesis to occur. Dephosphorylation of the tyrosine 15 on doubly phosphorylated Cdc2 by Cdc25 initiates its activity.

5.   START/restriction point is that time in G_1 after which cells are committed to entering the S phase even if inducers, such as nutrients, mitogens or growth factors, are removed.

MCAT/GRE-Style Questions

1. b; 2. d; 3. a; 4. a; 5. a; 6. d; 7. c; 8. b; 9. c; 10. d; 11. d

Chapter 14. Gene Control in Development

Testing Yourself on the Concepts

1.   A common strategy used to study the development of model organisms is to create a collection of mutants from that organism and screen the collection for mutants altered in development. For example, one can screen *Arabidopsis* mutants for altered flowers. When such a flower mutant is identified, it can be inferred that this mutant contains a mutation in a gene(s) required for flower development. The next step is to clone the mutation-defined gene and determine the pattern of mRNA expression and the distribution of the encoded proteins in the different spatial regions of the organism. This analysis determines in which parts of the organism the cloned gene functions.

Because all organisms are not amenable to such a genetic approach, investigators can start with a cloned gene known to affect the development of other organisms and find the homologue in a new organism. This is a common strategy used to study human development.

2. The drosophila embryo is a syncytium before the blastula stage. Early patterning events occur before the blastula stage that are dependent on the diffusion of RNAs and transcription factors that ultimately result in gradients of morphogens. The mature myotube is also a syncytium. In both of these cells, transcription and translation are not compartmentalized such that factors can freely diffuse to the site of action.

3. MCM1 is a factor involved in the mating-type conversion in yeast, binds to the P box in the URSs, and increases transcription of a-specific genes. MEFs are proteins involved in skeletal muscle development that bind to a muscle-specific enhancer and interact with MRFs to promote muscle gene transcription. APETELA-1 (an A-class gene) and AGAMOUS (a C-class gene) are MADS box proteins that may control transcription of flower genes.

4. A and B double mutants: a flower with four whorls consisting of carpels, carpels, carpels, carpels. B and C double mutant: a flower with four whorls consisting of sepals, sepals, sepals, sepals. A and C double mutant: leaves, petal/stamens hybrid, petals/stamens hybrid, new primordia (this is a result of loss of C function that results in a reiterative structure).

5. In yeast, the SIN gene products, which are part of the chromatin, repress HO transcription and mating-type conversion, perhaps by maintaining the HO regulatory region in a configuration that prevents binding of transcription factors. In mammals, MyoD, Myf5, and myogenin are able to remodel chromatin, which allows transcription factors access to muscle genes. In drosophila, Hox gene expression is controlled by two classes of genes that modulate chromatin structure: the trithorax group and the polycomb group. Polycomb proteins bind to multiple chromosomal locations and inactivate chromatin, which keeps developmental hox genes transcriptionally silent. Trithorax proteins are necessary for keeping many homeotic genes transriptionally active.

6. Id functions to prevent MyoD-E2A heterodimerization, which is required for transcriptional activation of muscle genes and muscle development. Id prevents this heterodimerization because it binds to MyoD and E2A, preventing their interaction with each other. EMC is analogous in function to Id in that it binds Ac and Sc proteins, thus inhibiting their association with Da and transcriptional activation of neural genes.

MCAT/GRE-Style Questions

1. a; 2. c; 3. b; 4. a; 5. a; 6. b; 7. d; 8. b; 9. b; 10. d; 11. a; 12. b; 13. d; 14. d

Chapter 15. Transport across Cell Membranes

Testing Yourself on the Concepts

1. The K_m should be about 1 mM. At a K_m of 10^{-6} M, the glucose transporter would be fully saturated with bound glucose over the entire concentration range of 3 to 7 mM blood glucose. With a K_m of 1 mM, comparatively little decrease in glucose uptake occurs with fasting, while with feasting some increase in the rate of glucose uptake is still possible. The K_m for GLUT1 glucose transporter in the erythrocyte membrane is 1.5 mM.

2. As lipids, steroid hormones would be expected to readily permeate and dissolve into biological membranes. They are, therefore, capable of interacting with an intracellular receptor protein, although they could also interact with a cell-surface receptor. In nature, steroid hormone receptor proteins are cytosolic, water-soluble proteins. In contrast, membrane-impermeant hormones must interact with a cell-surface receptor. They have no access to an intracellular receptor protein.

3.   Membrane transport proteins typically transport hydrophilic substances across biological membranes. Transmembrane helices of the transport proteins can associate to define an aqueous domain in which the transported substance can be moved while never being exposed to the hydrophobic environment of the membrane lipids.

4.   Although glucose movement is energetically unfavorable, the movement of Na^+ into cells is energetically favorable. The concentration of Na^+ inside a cell is low relative to that outside. The sum of the two transport processes mediated by the symport has a net $\Delta G < 0$, an energetically favorable situation. The Na^+ concentration gradient is maintained by the Na^+/K^+ ATPase, which utilizes ATP as an energy source.

5.   Water influx is important in stomata opening. Rapid rates of water entry require the presence of water-channel proteins in the plant plasma membrane.

MCAT/GRE-Style Questions

1. d; 2. a; 3. b; 4. a; 5. a; 6. d; 7. a; 8. b; 9. b; 10. d; 11. b; 12. d; 13. a

Chapter 16. Cellular Energetics: Glycolysis, Aerobic Oxydation, and Photosynthesis

Testing Yourself on the Concepts

1.   The pmf is generated by a voltage and chemical (proton) gradient across the inner membrane of mitochondria and the thylakoid membrane of chloroplasts. Like ATP, the pmf is a form of stored energy, and the energy stored in the pmf may be converted to ATP by the action of ATP synthase.

2.   In addition to providing energy to power ATP synthesis, the pmf also provides the energy used by several active transport proteins to move substrates into the mitochondria and products out of the mitochondria. The OH^- gradient, which results from generation of the pmf by electron transport, is used to move HPO_4^{2-} into the matrix, and the voltage gradient contribution of the pmf favors exchange of ADP for ATP.

3. O_2-generating photosynthesis uses the energy of absorbed light to create, via electron donation to quinone, the powerful oxidant P^+ form of the reaction center chlorophyll. This in turn acts to remove electrons from H_2O, a poor electron donor. The electrons are then passed along an electron transport chain, and the stored energy is converted to other forms for subsequent use in ATP synthesis and carbon fixation. The O_2 is not used in subsequent reactions in this pathway and thus is a by-product of the removal of electrons from H_2O.

4.   The Calvin cycle reactions are thought to be inactivated in the dark to conserve ATP that is needed for the synthesis of other cell molecules. The mechanism of inactivation depends on the enzyme; examples include pH-dependent and Mg^{2+}-dependent enzyme regulation, as well as reversible reduction-oxidation of disulfide bonds within certain Calvin cycle enzymes.

MCAT/GRE-Style Questions

1. c; 2. b; 3. d; 4. d; 5. b; 6. b; 7. d; 8. b; 9. c; 10. b; 11. d; 12. c; 13. d; 14. a

Chapter 17. Protein Sorting: Organelle Biogenesis and Protein Secretion

Testing Yourself on the Concepts

1. In the absence of any signal or targeting sequence, proteins stay in the compartment of their initial synthesis, be it the cytosol, mitochondrial matrix, or chloroplast stroma. It should be noted that low molecular proteins made in the cytosol can diffuse through nuclear pores into the nucleus without any requirement for a signal or targeting sequence.

2. Protein import into chloroplasts, mitochondria, and peroxisomes is post-translational, while import into the ER is cotranslational. Any involvement of chaperone class proteins during import into the ER is on the lumenal side of the ER. Chaperones (hsc 70) have an active cotranslational role in import into the ER in yeast but not in animal cells. Protein import into chloroplasts and mitochondria has a role for chaperones in the cytosol in maintaining the protein in an elongated form during import and within the lumen of the organelle in perhaps helping to power translocation as well as aiding in proper protein folding. Peroxisomal proteins are imported in a folded state, and chaperone class proteins have no direct role in their import.

3. Such evidence is summarized in the section entitled "Polypeptides Move through the Translocon into the ER Lumen" (p. 699). One of the key pieces of experimental evidence was that the nascent polypeptide chain could be chemically cross-linked to TRAM and the Sec61 complex. Through a series of experiments, both genetic and biochemical, the Sec61 complex was shown to form the actual translocon channel.

4. Because the protein contains disulfide bonds, it must have been translocated from the cytosol into the ER where the disulfide bonds would form. As the protein contains no hydrophobic segments typical of transmembrane α helices, barring the possibility that the protein is GPI anchored, the protein is not likely to be a membrane protein. If the protein is GPI anchored, it is likely to be on the cell surface. Otherwise, based on the available data, it is a soluble protein that may reside either within an organelle of the secretory pathway (ER, Golgi, transport vesicle, or lysosome) or have been secreted into the extracellular fluid.

5. Carbohydrate groups may be antigenic and hence provoke an immune response; this can have serious health consequences. If proteins contain certain exposed carbohydrates, such as galactose or mannose, they may be rapidly cleared from the bloodstream by binding to sugar-specific endocytic receptors. The protein may also fail to fold properly in the absence of normal or complete glycosylation and hence may be subject to quality control in the secretory pathway.

6. The vesicles are envisioned as carrying both membrane components and soluble lumenal proteins in a retrograde direction within the Golgi stack—from the trans- to the medial- and from the medial- to the cis-Golgi and also from the cis-Golgi back to the rough ER.

7. The likely site of entry is an acidic endosome. The cell surface is normally at a neutral pH. Only if the virus was internalized and entered into an acidic endosome would the HA fusion protein be acid pH activated.

MCAT/GRE-Style Questions

1. b; 2. c; 3. a; 4. d; 5. d; 6. c; 7. b; 8. d; 9. a; 10. c; 11. d; 12. b; 13. a; 14. c

Chapter 18. Cell Motility and Shape I: Microfilaments

Testing Yourself on the Concepts

1.   Actin filaments may contribute to cell movement through actin polymerization or by serving as tracks upon which other cellular components may be carried. Actin polymerization is capable of generating force to produce movement and is important for extending the leading edge of moving cells, the acrosome reaction in echinoderm sperm, and the intracellular movements of certain infectious bacteria and viruses. Actin filaments also support movement in nonmuscle cells by acting as "highways" along which various myosin motors may transport vesicles and other actin filaments. Examples of this type of movement include vesicular trafficking during secretion and axonal transport, cytoplasmic streaming in algae, contraction of the circumferential belt to alter cell shape, contractile ring contraction during cytokinesis, regulation of cortical tension, and cortical contraction during cell movement.

2.   Actin filaments of identical structure and composition may be involved in different processes within the same cytoplasm due to the action of the large number of proteins that interact with actin filaments. These proteins include cross-linking proteins, membrane-binding proteins, myosin motor proteins, assembly-regulating proteins, filament-severing proteins, and filament-capping proteins.

3.   All types of myosin are composed of one or two heavy chains and several light chains. Each heavy chain contains a related head domain capable of using energy from ATP hydrolysis to move toward the plus end of an actin filament. For each myosin type, different light chains may be present, but all associate with the neck region just adjacent to the head domain. The different myosin types are primarily distinguished by variable tail domains, which specify the particular cellular component that a given myosin recognizes and thus moves relative to actin filaments.

4.   In skeletal muscle, Ca^{2+}-dependent contraction depends on four proteins, tropomyosin and troponins C, I, and T, which associate with thin filaments. Troponin C binds Ca^{2+} and, depending on the cytosolic Ca^{2+} concentration, positions tropomyosin, via troponin T and I, on the thin filament to either block (if Ca^{2+} is low) or expose (if Ca^{2+} is high) the myosin-binding sites on the thin filament. Thus, when Ca^{2+} is high, myosin in the thick filament can access and walk along thin filaments and thereby produce contraction. In smooth muscle, Ca^{2+} may regulate contraction by one of two mechanisms. The first is similar to that described for skeletal muscle except that the protein caldesmon performs a functionally similar role to the three troponin proteins. Caldesmon action also may also be influenced through phosphorylation by protein kinase C. The second mechanism involves regulation by Ca^{2+} of myosin activity. This may occur either through direct binding of Ca^{2+} to myosin regulatory light chains or Ca^{2+}-dependent phosphorylation of myosin regulatory light chains. In either case, a rise in cytosolic Ca^{2+} serves to trigger the light chains to activate myosin movement.

MCAT/GRE-Style Questions

1. d; 2. b; 3. b; 4. b; 5. d; 6. b; 7. a; 8. b; 9. c; 10. d; 11. b; 12. d; 13. c; 14. d; 15. c

Chapter 19. Cell Motility and Shape II: Microtubules and Intermediate Filaments

Testing Yourself on the Concepts

1. The basis of microtubule polarity is the head-to-tail assembly of α, β-tubulin heterodimers, which results in a crown of α-tubulin at the (−) end and a crown of β-tubulin at the (+) end. In nonpolarized animal cells, (−) ends are typically associated with MTOCs, and (+) ends may extend toward the cell periphery. Other arrangements occur in different types of cells, but the (−) ends are associated with a MTOC in most cases. Microtubule motors can "read" the polarity of microtubule, and a specific motor protein will transport its cargo toward either the (+) or the (−) end of the microtubule.

2. The best understood proteins involved in regulating microtubule assembly are the assembly MAPs. These proteins, divided into Type I and Type II MAPs, bind to microtubules and promote assembly, increase microtubule stability, and in some cases cross-link microtubules into bundles. Other proteins that regulate assembly include katinin, which acts to sever microtubules, and Op 18, which promotes the frequency of microtubule catastrophe.

3. The appendages (cilia and flagella) used for cell swimming contain a highly organized core of microtubules and associated proteins. This core, termed the axoneme, is typically made of nine outer doublet microtubules and two central pair microtubules (termed the 9 + 2 arrangement). Each outer doublet consists of a 13-protofilament microtubule and a 10-protofilament microtubule, while the central pair of microtubules each contain 13 protofilaments. Cell movement depends on axoneme bending, which in turn depends on force generated by axonemal dyneins. These motor proteins act to slide outer doublet microtubules past each other, but this sliding motion is converted into bending due to restrictions imposed by cross-linking proteins in the axoneme, and perhaps by the action of inner arm dyneins.

4. The first step in the assembly of a bipolar spindle is the separation of duplicated centrosomes. Microtubule motors align and slide overlapping, oppositely oriented microtubules that originate from each centrosome. Alignment may involve a (−) end-directed KRP, while sliding appears to involve a (+) end-directed KRP. Sliding apart of centrosomes may also be aided by cytoplasmic dynein at the cell cortex, which could act to "reel in" astral microtubules. After formation of the bipolar spindle, dynein and/or KRPs near the (−) ends of spindle microtubules act to organize the spindle poles and retain pole attachment to centrosomes. Microtubule motors on kinetochores are involved in capturing microtubule (+) ends emanating from spindle poles, in developing stable attachments (via kinetochore microtubules) between the chromosomes and spindle poles, and in positioning the chromosomes at the metaphase plate. During anaphase A, the role of motor proteins at the kinetochore is controversial. There is evidence that poleward chromosome movement does not require ATP but that motor proteins may be involved in maintaining kinetochore attachment to depolymerizing microtubules. During anaphase B, the spindle poles are moved apart by sliding produced by (+) end-directed motors on the overlapping polar microtubules and/or pulling of astral microtubules by (−) end-directed motors at the cell cortex.

5. Disassembly of several types of intermediate filament (keratin, vimentin, lamin) can be induced by phosphorylation of a serine residue in the N-terminal domain. This typically occurs early in mitosis and the phosphorylation reaction may be catalyzed by cdc2 kinase. Disassembly of the cytoplasmic intermediate filament arrays probably facilitates the cytoplasmic rearrangement involved in mitosis and cytokinesis, while disassembly of the nuclear lamina contributes to the disassembly of the nuclear envelope. Phosphatase-dependent removal of the phosphate group permits filament reassembly as cells leave mitosis.

1. a; 2. d; 3. d; 4. b; 5. a; 6. d; 7. c; 8. d; 9. b; 10. d; 11. c; 12. c; 13. d; 14. a

Chapter 20. Cell-to-Cell Signaling: Hormones and Receptors

Testing Yourself on the Concepts

1. Ligand binding to G protein-coupled receptors activates the associated trimeric G protein, which in turn activates adenylyl cyclase to generate the intracellular second messenger, cAMP. Binding of ligand to the receptor results in a change in the conformation of the receptor. This altered receptor binds to the trimeric G protein in such a way that GDP is displaced from the inactive GDP-$G_{s\alpha}$ complex. Subsequent binding of GTP to $G_{s\alpha}$ produces an activated G protein. This activated GTP-$G_{s\alpha}$ complex dissociates from the $G_{\beta\gamma}$ complex and binds to and activates adenylyl cyclase. Adenylyl cyclase then converts ATP to cAMP. Activation of adenylyl cyclase is short-lived because of the intrinsic GTPase activity in G proteins. Hydrolysis of the GTP bound to $G_{s\alpha}$ to GDP leads to reassociation of $G_{s\alpha}$ with $G_{\beta\gamma}$ and inactivation of adenylyl cyclase. See Figure 20-16.

2. Receptor tyrosine kinases (RTKs) contain an extracellular ligand-binding domain, a single hydrophobic transmembrane domain, and a cytosolic domain that includes a region with protein kinase activity. Binding of ligand to the receptor causes many RTKs to dimerize. The protein kinase of each receptor monomer then phosphorylates specific tyrosine residues in the cytosolic domain of its dimer partner in a process called autophosphorylation. The first step in autophosphorylation is the phosphorylation of tyrosines in a region near the catalytic site known as the phosphorylation lip. This facilitates binding of ATP or protein substrates. The second step involves phosphorylation of other cytosolic sites, which serve as docking sites for other proteins. Cytosolic phosphotyrosines interact with the adapter protein, GRB2, through SH2 domains. GRB2 interacts with Sos, which in turn interacts with Ras. Ras is an intracellular GTPase switch protein that cycles between an inactive GDP-bound form and an active GTP-bound form. Ras cycling requires the assistance of a guanine nucleotide exchange factor (GEF) and a GTPase-activating protein (GAP). Binding of Sos, a GEF, to inactive Ras causes a conformational change that results in the release of GDP and binding of GTP. The activated Ras then induces a kinase cascade that culminates in activation of MAP kinase. See Figures 20-22 and 20-23.

3. Binding of ligands to certain G protein–coupled receptors (GCPRs) and receptor tyrosine kinases (RTKs) activates the membrane-associated enzyme phospholipase C (PLC). Cleavage of the membrane-bound phosphoinositide (PIP_2) by PLC produces 1,2-diacylglycerol (DAG), a lipophilic molecule that remains linked to the membrane, and free inositol 1,4,5-trisphosphate (IP_3), which can diffuse into the cytosol. IP_3 binds to a Ca^{2+}-channel protein and induces the opening of the channel allowing Ca^{2+} ions to exit from the endoplasmic reticulum (ER) into the cytosol. The rise in cytosolic Ca^{2+} level causes protein kinase C to bind to the cytosolic side of the plasma membrane, where it can be activated by the membrane associated DAG. The activated protein kinase C then mediates a varied array of cellular responses. A sustained rise in cytosolic Ca^{2+} mediated by IP_3 also requires opening of plasma-membrane Ca^{2+} channels, called store-operated channels (SOCs). The rise in Ca^{2+} concentration is transient because Ca^{2+}-ATPases located in the plasma membrane and ER membrane actively pump Ca^{2+} from the cytosol to the cell exterior and ER lumen, respectively. Also, IP_3 is rapidly hydrolyzed to 1,4-bisphosphate, which does not stimulate Ca^{2+} release from the ER. See Figure 20-39.

MCAT/GRE-Style Questions

1. b; 2. c; 3. a; 4. c; 5. b; 6. a; 7. d; 8. c; 9. a; 10. d; 11. b; 12. c; 13. d

Chapter 21. Nerve Cells

Testing Yourself on the Concepts

1.   Neurons conduct signals along axons by the unidirectional propagation of a transient electrical disturbance called the action potential. The action potential depends on the sequential opening of voltage-gated Na^+ and K^+ channels to first depolarize and then repolarize a localized region of the plasma membrane. The directional nature of the action potential is possible due to the timed closing of voltage-gated Na^+ channels. When the action potential reaches the end of the axon, two possible mechanisms are possible depending on the relationship between the presynaptic and postsynaptic cell. If an electrical synapse connects the two cells, the action potential from the presynaptic cell travels as a membrane depolarization as ions move through gap junctions to the postsynaptic cell. If a chemical synapse connects the two cells, the presynaptic cell converts the electrical signal (action potential) into a chemical signal through the triggered release of neurotransmitters into the synaptic cleft. The neurotransmitters are recognized by receptors on postsynaptic cell, and the chemical signal is converted back into an electrical signal by the opening of gated ion channels in the postsynaptic cell.

2.   Myelin is a region of specialized plasma membrane derived from glial cells that wraps around vertebrate axons. Myelin prevents movement of ions between the axonal cytosol and the extracellular fluid but does not completely sheath the entire length of the axon. Nonmyelinated regions, called nodes, are present at intervals along the axon, and the proteins involved in controlling ion movement across the membrane are clustered in these nodes. In myelinated axons, the action potential essentially jumps from one node to the next because the large number of ions associated with depolarization at one node can spread rapidly and without attenuation to the next node. Myelination thus provides more rapid movement of the action potential and also reduces the diameter of the neuron required to transmit an action potential at a given rate. Loss of myelin and subsequent reduction in the rate of action potential propagation is associated with neurological diseases such as multiple sclerosis.

3.   Acetylcholine may be excitatory or inhibitory because there is more than one type of acetylcholine receptor and downstream signal cascade. Acetylcholine is excitatory when nicotinic acetylcholine receptors are present on the postsynaptic cell. These receptors are ligand-gated Na^+ and K^+ channels, and binding opens the channels, allowing an influx of positive ions that produces membrane depolarization leading to generation of an action potential. The stimulation of skeletal muscle contraction is an example of this pathway. In comparison, acetylcholine is inhibitory when muscarinic acetylcholine receptors are present on the postsynaptic cell. Here, binding of acetylcholine to a G protein-coupled receptor activates a cascade leading to the opening of K^+ channels and hyperpolarization of the membrane, making an action potential less likely to form. Reducing the rate of heart muscle contraction is an example of this pathway.

4.   Light detection requires a receptor that can absorb a photon of light, while odorant detection involves receptors specific for a given odor. There are four types of light receptors and approximately 1,000 types of odorant receptors. However, in both cases, the receptors are members of the G protein-lined family of receptor, so that detection of the stimulus leads to activation of a G protein-regulated signal cascade. In the dark, the cells involved in light detection (rods and cones) contain high concentrations of the second messenger cGMP, which acts to keep ligand-gated Na^+ channels open. Thus these cells are depolarized and continually secreting neurotransmitters to postsynaptic cells in the absence of light. Detection of photons activates the G protein pathway, which activates destruction of cGMP, closing of the Na^+ channels, hyperpolarization of the membrane, and termination of signaling to the postsynaptic cell, which is interpreted as light. In comparison, odorant receptors at rest have a similar membrane potential to that of most neurons and are therefore not sending signals to postsynaptic cells in the absence of the odorant. Upon binding of an odorant to the receptor, the coupled G protein activates

production of a second messenger (cAMP), which binds and opens ligand-gated cation channels in the plasma membrane. The influx of positive ions into the cell depolarizes the membrane, leading to an action potential and eventual transmission of the signal to the postsynaptic cell for eventual interpretation as a specific odor.

MCAT/GRE-Style Questions

1. d; 2. d; 3. a; 4. c; 5. b; 6. c; 7. b; 8. b; 9. a; 10. d; 11. a; 12. c; 13. c; 14. b

Chapter 22. Integrating Cells into Tissues

Testing Yourself on the Concepts

1. Homophilic, like-binds-like adhesion occurs between cells of a single type, and heterophilic adhesion occurs between cells of different types. Two sets of interactions, head-to-head and side-to-side, are postulated to cause the clustering or "zippering" of cadherins in specialized adhesion junctions (e.g., adherens junctions and desmosomes). Side-to-side interactions are necessary for cadherin dimer formation. They may also occur in cadherin clustering. Head-to-head interactions occur between cadherin dimers in adjacent cell membranes.

2. Cytosolic Ca^{2+} concentrations are low, less than 10^{-6} M, while extracellular calcium concentrations are typically millimolar. Elevation of Ca^{2+} levels produces a conformational change in gap junctions and the channel closes. This prevents the leakage of small molecule cellular contents from adjacent cells when one cell in the interconnected layer is damaged.

3. Collagen is a glycoprotein, while cellulose is a sugar (glucose) polymer. Collagen synthesis begins in the ER as with other secreted proteins and proceeds through a set of secondary modifications and polymerization reactions (trimer formation) in the ER and then further modifications within the Golgi apparatus. N- and C-terminal peptides are removed so that large-scale polymerization of collagen occurs extracellularly. Collagen crosslinking is extracellular. Cellulose is synthesized from glucose by a cell-surface enzyme. All synthetic steps occur at the cell surface, and cellulose synthesis and polymerization is entirely an extracellular process.

4. Proteoglycans contain a very high ratio of carbohydrate to protein. The basic structure of a proteoglycan consists of a core protein modified by several three-sugar "linkers" to which glycosaminoglycans (GAGs) such as heparan sulfate attach. GAGs are long-repeating linear polymers of specific disaccharides. The number, length, and composition of the GAG chains attached to each core protein may vary. Such a structure is exemplified by aggrecan, in which the proteoglycan units are aggregated through the core protein to a long molecule of the saccharide hyaluronan. The aggregation to hyaluran is aggrecan specific. This gives a macromolecule that occupies a large volume in a gel-like manner that resists deformation. This is essential for distributing load in weight-bearing joints.

5. Cells are attached to ECM, and this attachment may impede cell migration. Protease activity, for example, fibrinogen and matrix-specific metalloproteinases (MMPs), by degrading matrix components loosens the ECM and thus facilitates cell migration.

MCAT/GRE-Style Questions

1. c; 2. d; 3. b; 4. a; 5. a; 6. d; 7. b; 8. a; 9. c; 10. c; 11. a; 12. b; 13. d

Chapter 23. Cell Interactions in Development

Testing Yourself on the Concepts

1. In principle, an inductive signal may act either through a gradient or by inducing a cascade of signals (relay model). Experiments with chick neural-tube explants show that Hh acts through a gradient. When one of four different concentrations of Hh was added to the explant, one of four different cell types was produced. At each concentration a different cell type was produced. These data strongly suggest that different cell types are formed in vivo in response to a ventral-to-dorsal gradient in Hh concentration. The precise mechanisms by which the concentration of Hh specifies different fates is not known. Studies in several other systems, however, may provide clues to how cells respond to different concentrations of a morphogen. For instance, experiments in *Xenopus* using the TGFβ family member activin indicate that cells respond to the number of receptors occupied by ligand, rather than the fraction of receptors occupied. Ultimately, the number of receptors occupied must be translated into changes in the pattern of gene expression. While little is known about the mechanisms by which different concentrations of Hh influence gene expression, considerable progress has been made in dissecting the mechanisms by which *Drosophila* cells respond to the graded expression of Spaetzle, a ligand for the Toll receptor controlling early stages of patterning along the dorsal (low) to ventral (high) axis. The expression of specific genes is determined by differences in the *cis*-acting control regions. For instance, different target genes contain binding sites for Dorsal of different affinity; high-affinity sites will support expression of target genes at low Dorsal concentration, while low-affinity sites support Dorsal-dependent expression only at high concentration. The expression of specific genes also may reflect the number of *cis*-acting regulatory sites and the function of additional spatially restricted transcription factors.

2. The inductive activity of Hh is contained in an N-terminal-20-kDa polypeptide fragment derived from full-length Hh (45 kDa) by an endoproteolytic mechanism. Endoproteolysis leads to the attachment of cholesterol to the C-terminus of the 20-kDa fragment; due to the hydrophobic character of cholesterol, Hh is tethered to the extracellular leaflet of the plasma membrane. Membrane attachment through cholesterol appears to be used by only a small number of other proteins. The net effect is to anchor the inductive fragment to the cell surface, thereby limiting the range of its inductive activity.

3. Three proteins—CED-9, CED-4, and CED-3—play key roles in regulating apoptosis during normal development of *C. elegans*. They function as a regulator, adapter, and effector, respectively. The pro-apoptotic function of CED-4 is directly suppressed by the anti-apoptotic function of CED-9. CED-3 is a caspase, a cysteine protease that selectively cleaves proteins at sites just C-terminal to aspartate residues. CED-3 is activated by CED-4. Loss-of-function mutations in *ced-9* lead to the death of all cells. In contrast, loss-of-function mutations in *ced-3* and *ced-4* result in the survival of cells that normally die through programmed cell death.

4. It is thought that in many cases intermediate targets produce signals that attract growth cones toward them; growth cones migrate up a gradient of an attractant toward the target. Once growth cones reach the intermediate target they must then grow down the gradient, from a high to a low concentration of attractant. Studies on netrins suggest a possible mechanisms to account for this feature of growth cone behavior. While a gradient of netrin attracts axons from retinal ganglion cells in tissue culture, growth cones incubated with an inhibitor of cAMP-dependent protein kinase (cAPK) are repelled by netrin. Based on this observation the following model can be proposed: Upon reaching an intermediate target, levels of cAMP in the growth cone are decreased leading to conversion of an attractant response to a repellent. Hence, the signal that initially served as an attractant is converted into a repellent by virtue of a change in the intracellular signaling pathways acting in the growth cone.

1. d; 2. a; 3. c; 4. a; 5. b; 6. c; 7. a; 8. a; 9. d; 10. a; 11. c; 12. b

Chapter 24. Cancer

Testing Yourself on the Concepts

1. Cell-surface receptors respond to extracellular growth signals and transmit this information, by signal-transduction pathways, to intracellular targets. Alteration of either of these components can result in oncogenesis. Any one of a number of changes in receptors (mutation, overexpression, inappropriate subcellular localization) can cause constitutive, rather than "on/off," sensitivity to ligand. Similar changes in components of the signal-transduction cascade can occur. A common result of all of these alterations is unregulated activity of protein tyrosine kinases or phosphorylation of novel substrates.

2. Gain-of-function mutations convert proto-oncogenes to oncogenes or cause overexpression of proto-oncogenes. These mutations are dominant, meaning that change in one allele is sufficient for the phenotype. In contrast, loss-of-function mutations are recessive, so that both alleles must be altered for the phenotype to be observed. Loss-of-function mutations occur in tumor-suppressor genes, preventing their ability to, for example, establish checkpoints, repair DNA, inhibit cell proliferation, or promote apoptosis. An exception is the tumor-suppressor gene *p53*, for which a single mutation is sufficient for a gain of function.

3. The current understanding of the genetic basis of the development of cancer focuses on genes that regulate cell growth and division. The "coin" is the gene. In the wild-type state, growth and division are tightly regulated; in the mutant state, the protein produced by the gene is defective, produced at a higher level, or has a greater activity than normal. Any of these changes can derange the checks and balances in the cell's physiology.

4. The increased incidence of cancer with age is explained by a "multi-hit" model, with successive mutations or alterations in gene expression corresponding to the discrete stages leading to a lethal tumor. For example, many colon cancers contain mutations in *APC, DCC* and *p53*, tumor-suppressor genes, and in *ras*. The *APC* mutation that causes overexpression of *myc* is found in polyps, an early stage of colon cancer, while *p53* mutation is required for malignancy. In mice, overexpression of *myc* or expression of ras^D cause cancer only after a long lag. However, these two genes act synergistically to cause cancer in at least one-third the time of either one alone.

MCAT/GRE-Style Questions

1. b; 2. c; 3. d; 4. c; 5. b; 6. c; 7. c; 8. d; 9. c; 10. d; 11. d